Lecture Notes in Computer Science   8506

Commenced Publication in 1973
Founding and Former Series Editors:
Gerhard Goos, Juris Hartmanis, and Jan van Leeuwen

Hua Wang   Mohamed A. Sharaf (Eds.)

Databases Theory and Applications

25th Australasian Database Conference, ADC 2014
Brisbane, QLD, Australia, July 14-16, 2014
Proceedings

 Springer

Volume Editors

Hua Wang
Victoria University
College of Engineering and Science
Centre for Applied Informatics (CAI)
Ballarat Road
Footscray, VIC 8001, Australia
E-mail: hua.wang@vu.edu.au

Mohamed A. Sharaf
The University of Queensland
Faculty of Engineering, Architecture and Information Technology
School of Information Technology and Electrical Engineering
Brisbane St. Lucia, QLD 4072, Australia
E-mail: m.sharaf@uq.edu.au

ISSN 0302-9743   e-ISSN 1611-3349
ISBN 978-3-319-08607-1   e-ISBN 978-3-319-08608-8
DOI 10.1007/978-3-319-08608-8
Springer Cham Heidelberg New York Dordrecht London

Library of Congress Control Number: 2014941783

LNCS Sublibrary: SL 3 – Information Systems and Application,
incl. Internet/Web and HCI

Typesetting: Camera-ready by author, data conversion by Scientific Publishing Services, Chennai, India

Printed on acid-free paper

Springer is part of Springer Science+Business Media (www.springer.com)

Preface

It is our pleasure to present to you the proceedings of the 25th Australasian Database Conference (ADC2014), which took place in Brisbane, Australia. ADC2014 is an annual forum for researchers and practitioners from Australia, New Zealand and around the world to share the latest research progress and novel applications of database systems, data driven applications, and data analytics. The mission of ADC is to exchange novel research solutions to problems of today's information society that fulfill the needs of heterogeneous applications and environments, as well as to identify new issues and directions for future research and development work. ADC2014 seeks papers from academia and industry presenting research on all practical and theoretical aspects of advanced database theory and applications, as well as case studies and implementation experiences. All topics related to databases are of interest and within the scope of the conference. ADC gives researchers and practitioners a unique opportunity to share their perspectives with others interested in the various aspects of database systems.

The ADC 2014 Program Committee accepted those papers considered to be of ADC quality without setting any predefined quota, and was impressed by the quality of the submissions. The conference received 38 submissions and accepted 15 papers. The Program Committee who selected the papers consisted of 28 members from around the globe including Singapore, Bangladesh, Germany, Ireland, Japan, Switzerland, China, and the United States. The Program Committee was thorough and dedicated to the reviewing process with each paper peer reviewed in full by at least two independent reviewers, and in some cases three or four referees produced independent reviews. A conscious decision was made to select papers for which all reviews were positive and favorable. While this challenged the determination, and some high-quality papers were finally not included, we are confident that the results demonstrate a very solid program and that each paper makes a strong contribution to the proceedings.

We would like to thank all our colleagues who served on the Program Committee or acted as external reviewers. We would also like to thank all the authors who submitted papers, both accepted and rejected, as well as the conference attendees. This conference is held for you, and we hope that these proceedings provide an overview of our vibrant research community and its activities. We encourage all database researchers to contribute and make submissions to the next ADC conference.

July 2014

Hua Wang
Mohamed A. Sharaf

General Chair's Welcome Message

Welcome to the 25th Australasian Database Conference (ADC2014)! ADC is a leading Australia and New Zealand based international conference on research, development and applications of database systems and related areas. Previous ADC conferences were held as part of the Australasian Computer Science Week (ACSW). In the past 10 years, ADC was held in Adelaide (2013), Melbourne (2012), Perth (2011), Brisbane (2010), Wellington (2009), Wollongong (2008), Ballarat (2007), Hobart (2006), Newcastle (2005), and Dunedin (2004).

Australasia has an increasingly large, very active and internationally highly visible group of database researchers. Based on wide community consultation, ADC 2014 departs from its tradition as part of the Australasian Computer Science Week (ACSW) to become an independent conference with an expanded research program, a PhD School and a community-building focus. So, in that sense, ADC 2014 is the first of the new ADC conference series.

The conference this year had two eminent keynote speakers: Timos Sellis from the Royal Melbourne Institute of Technology, Australia, and Divesh Srivastava from AT&T, USA. In addition to 15 full papers and 6 short papers carefully selected by the Program Committee, we were also very fortunate to have three invited talks presented by world-leading researchers, ChengXiang Zhai from UIUC, Lei Chen from HKUST and Jeffrey Yu from CUHK. We also have a three-day PhD School program as part of this year's ADC. We wish to take this opportunity to thank all speakers, authors and organisers. I would also specially thank the Program Committee co-chairs Hua Wang and Mohamed A. Sharaf for their dedication and effort in ensuring a high quality program. I would also like to thank our PhD School convenor, Heng Tao Shen, and our key local organiser, Kath Williamson, for their contributions to making this year's new ADC a success.

As a member of the new ADC's Steering Committee, I would like to sincerely thank my fellow Committee members: Rao Kotagiri (The University of Melbourne), Timos Sellis (RMIT), Gill Dobbie (University of Auckland), Alan Fekete (The University of Sydney), Xuemin Lin (UNSW), and Yanchun Zhang (Victoria University), for their support to set a new direction for ADC and offering the opportunity to host the first new ADC conference at The University of Queensland in Brisbane.

Brisbane is a beautiful city and UQ has one of the best university campuses in Australia. All ADC2014 participants are sure to enjoy the conference, the campus and the city.

General Chair   Xiaofang Zhou (The University of Queensland)

Organization

General Chair

Xiaofang Zhou   University of Queensland, Australia

PC Co-chairs

Hua Wang   Victoria University, Australia
Mohamed A. Sharaf   University of Queensland, Australia

Steering Committee

Rao Kotagiri   University of Melbourne, Australia
Timos Sellis   RMIT University, Australia
Gill Dobbie   University of Auckland, New Zealand
Alan Fekete   University of Sydney, Australia
Xuemin Lin   University of New South Wales, Australia
Yanchun Zhang   Victoria University, Australia

Program Committee

Ying Zhang   University of New South Wales, Australia
Sebastian Maneth   University of Edinburgh, UK
Junhu Wang   Griffith University, Australia
Gang Li   Deakin University, Australia
Mohammed Eunus Ali   University of Engineering and Technology, Bangladesh
Michael E. Houle   National Institute of Informatics, Japan
Ruixuan Li   Huazhong University of Science and Technology, China
Sarana Nutanong   Johns Hopkins University, USA
Shichao Zhang   Guangxi Normal University, China
Panos Chrysanthis   University of Pittsburgh, USA
Chaoyi Pang   CSIRO, Australia
James A. Thom   RMIT University, Australia
Xue Li   University of Queensland, Australia
Yoshiharu Ishikawa   Nagoya University, Japan
Xiangmin Zhou   CSIRO, Australia

Markus Stumptner	University of South Australia, Australia
Annika Hinze	University of Waikato, New Zealand
Zahir Tari	RMIT University, Australia
Bing Tian Dai	Singapore Management University, Australia
Jinli Cao	Latrobe University, Australia
Miyuki Nakano	University of Tokyo, Japan
Evaggelia Pitoura	University of Ioannina, Greece
Bela Stantic	Griffith University, Australia
Hye-Young Paik	University of New South Wales, Australia
Ge Yu	Northeastern University, China
Laurianne Sitbon	Queensland University of Technology, Australia
Chengfei Liu	Swinburne University of Technology, Australia

Invited Talks

Analysing Big Trajectory Data: Theory, Algorithms and Applications

Kai Zheng

University of Queensland

Abstract. The prevalence of GPS sensors and mobile devices has enabled tracking the movements of almost any kind of moving objects such as vehicles, humans and animals. As a result, in the past decade we have witnessed unprecedented increase of trajectory data both in volume and variety. With some attributes such as variable lengths, uncontrolled quality, high redundancy and uncertainty and so on, trajectory data challenge the traditional methodologies and practices in many research areas including data storage and indexing, data mining and analytics, information retrieve, etc. Trajectory data management has been attracting numerous research interests from both academia and industry due to its tremendous value and benefits in a variety of critical applications like traffic analysis, fleet management, trip planning, location-based recommendation, etc. In this tutorial, we will talk about the challenges, techniques and open problems with the focus on similarity-based analytics, the foundation of trajectory management, and covering a range of topics from fundamental theory, algorithms to advanced applications.

Boosting Methods in Machine Learning

Chunhua Shen

University of Adelaide

Abstract. Many machine learning and data mining tasks favour fast and yet accurate classification methods. The classification speed is not only a matter of time-efficiency but is often crucial to achieve good accuracy. Standard kernel machines such as Support Vector Machine (SVM) are slow and methods for rapid classification have been pursued. Boosting classifiers have been so successful owing to its fast computation and yet comparable or sometimes better accuracy to kernel methods, being a standard method in many areas. Boosting as a representative ensemble learning method, which aggregates simple weak learners, can be seen as a flat tree structure when each learner is a decision-stump. When trees are used as weak learners, boosting methods learn a linearly weighted decision forest. We will overview the fundamental theory of boosting in the first part of this course.

Recently, structured learning has found many applications in text analysis and computer vision. Thus far it has not been clear how one can train a boosting model that is directly optimised for predicting multivariate or structured outputs. To bridge this gap, inspired by structured support vector machines, a boosting algorithm for structured output prediction is introduced, which we refer to as StructBoost. StructBoost supports nonlinear structured learning by combining a set of weak structured learners. As structured SVM generalises SVM, the StructBoost generalises standard boosting approaches such as AdaBoost, or LPBoost to structured learning. The resulting optimization problem of StructBoost is more challenging than Structured SVM in the sense that it may involve exponentially many variables and constraints. In contrast, for Structured SVM one usually has an exponential number of constraints and a cutting-plane method is used. In order to efficiently solve StructBoost, we formulate an equivalent 1-slack formulation and solve it using a combination of cutting planes and column generation. We show the versatility and usefulness of StructBoost on a range of problems.

Big Data Mining on SAP HANA

Hoyoung Jeung

SAP

Abstract. This talk will cover how the theories in data mining and machine learning are implemented and used in the state-of-the-art in-memory computing technology, SAP HANA. In particular, Dr.Jeung will share his extensive experience in predictive analysis on big business data, discussing about hidden insights when dealing with complex algorithms on extremely large data.

Statistical Methods for Mining Big Text Data

Chengxiang Zhai

University of Illinois Urbana-Champaign, USA

Abstract. Text data, broadly including all kinds of natural language text produced by humans (e.g., web pages, social media, email messages, news articles, government documents, and scientific literature), have been growing dramatically recently. This creates great opportunities for applying computational methods to mine large amounts of text data to discover all kinds of useful knowledge, especially knowledge about people's opinions, preferences, and behavior. Due to the difficulty in precisely understanding natural language by computers, scalable text mining algorithms tend to be based on statistical analysis and probabilistic reasoning. In this tutorial, I will systematically review the major statistical methods developed for mining text data, with a focus on covering probabilistic topic models for mining topics and topical patterns in text data, and statistical methods for integrating and analyzing scattered online opinions.

Crowdsourcing over Big Data, Are We There Yet?

Lei Chen

Hong Kong University of Science & Technology

Abstract. Recently, the popularity of crowdsourcing has brought a new opportunity to engage human intelligence into various data analysis tasks. Compared with computer systems, crowds are good at handling items with human-intrinsic values or features. Existing approaches develop sophisticated methods by utilizing the crowd as a new type of processor, a.k.a. HPU (Human Processing Unit). As a consequence, tasks executed on HPU are called HPU-based tasks. Now we are in the Big Data Era, a nature question arises: How about crowdsourcing over Big Data, are we there yet? In this talk, I will first briefly review the history of crowdsourcing and discuss the key issues related to crowdsourcing. Then, I will demonstrate the power of crowdsourcing in solving the well-known and very hard data integration problem, schema matching, and discuss how to migrate the power of crowdsourcing to a social media platform whose users can serve as a huge reservoir of workers. Finally, I will highlight some research challenges about crowdsourcing over Big Data.

Large Graph Processing

Jeffrey Yu

Chinese University of Hong Kong

Abstract. The real applications that need graph processing techniques to handle a large graph can be found from many real applications including online social networks, biological networks, ontology, transportation networks, etc. In this talk, we will discuss some selected research topics on graph mining and graph query processing over large graphs. For graph mining, we will focus on ranking nodes in a large graph. We will discuss ranking over trust networks, random-walk domination, and diversified ranking. For ranking nodes over trust network, we discuss how to take the trust score into consideration while ranking. For the random-walk domination, we discuss the techniques for handling item-placement in online social networks and ads-placement in advertisement networks. For diversified ranking, we discuss how to find top-k nodes that match the user query and are very different from each other. For graph query processing, we will discuss top-k structural diversity search, finding the maximal cliques in massive networks, and I/O efficient computing techniques that make a large directed graph small and simple. The other related topics may be also addressed in this talk.

Keynotes

Keynotes

Selecting Sources Wisely
for Integration

Xin Luna Dong[1], Theodoros Rekatsinas[2], Barna Saha[3],
and Divesh Srivastava[3]

[1]Google Inc., Mountain View, CA 94043, USA
lunadong@google.com
[2]University of Maryland, College Park, MD 20742, USA
thodrek@cs.umd.edu
[3]AT&T Labs-Research, Bedminster, NJ 07921, USA
{barna, divesh}@research.att.com

Abstract. Data integration is a challenging task due to the large numbers of autonomous data sources, which necessitates the development of techniques to reason about the costs and benefits of acquiring and integrating data. Too many sources can result in a huge integration cost, and low quality sources can be detrimental to the benefit of integration. In this talk, we present the problem of *source selection*, that is, identifying the subset of sources before integration that maximize the profit (benefit − cost) of integration, for static and dynamic sources. To address this problem, we propose techniques that, inspired by the marginalism principle in economic theory, integrate a source only if its marginal benefit is higher than its marginal cost. We quantify the integration benefit in terms of the quality of the integrated data, which is characterized using a set of data quality metrics, including coverage, freshness and accuracy, and develop statistical models for estimating these metrics. Although source selection is NP-complete, we show that for many practical cases solutions to our problem can be found in polynomial time with approximation guarantees. Finally, we empirically establish the effectiveness and scalability of our techniques on real-world and synthetic data.

Data Ecosystems: From Very Large Data Bases to Big Data Infrastructures

Timos Sellis

Computer Science & Info Tech
RMIT University
timos.sellis@rmit.edu.au

Abstract. Data ecosystems involve the coexistence of one or more data collections, typically databases, and their surrounding applications for data entry and retrieval. For decades, both data and ecosystem management have failed to address significant, costly and labor-consuming challenges which involve (a) the departure from databases focusing on alphanumeric data only, (b) their inability to be integrated and provide transparent access and composition facilities for heterogeneous data, (c) their static querying nature, which is deprived of personal, context-aware or interactive characteristics, (d) the enforcement of DBMS operation over monolithic servers, and, (e) the complete indifference to problems of evolution and adaptation over time.

In this talk we address issues around the methodologies, the theoretical and modeling foundations as well as the algorithmic techniques and the necessary software architectures that will facilitate the personalization, integration, and evolution management facilities for data ecosystems that operate over a decentralized infrastructure for a large variety of data types.

Table of Contents

Dynamic Sorted Neighborhood Indexing
for Real-Time Entity Resolution*

Banda Ramadan, Peter Christen, and Huizhi Liang

Research School of Computer Science, College of Engineering and Computer Science
The Australian National University, Canberra ACT 0200, Australia
{banda.ramadan,peter.christen,huizhi.liang}@anu.edu.au

Abstract. Real-time entity resolution is the process of matching query records in sub-second time with records in a database that represent the same real-world entity. Indexing techniques are used to efficiently extract a set of candidate records from the database that are similar to a query record, and that are then compared with the query record in more details. The sorted neighborhood indexing method, which sorts a database and compares records within a sliding window, has successfully been used for entity resolution of very large databases. However, because it is based on static sorted arrays, this technique is not suitable for dynamic databases. We propose a tree-based dynamic sorted neighborhood index that facilitates matching a stream of query records against a large and dynamic database in real-time. We evaluate our approach on two large data sets. Our results show that the times for both inserting and querying of records stays nearly constant as the index grows, and our approach achieves over one magnitude faster indexing and querying times compared to an earlier real-time entity resolution technique with comparable high matching accuracy.

Keywords: Dynamic indexing, data matching, braided tree.

1 Introduction

Massive amounts of data are being collected by most business and government organizations. Given that many of these organizations rely on information in their day-to-day operations, the quality of the collected data has a direct impact on the quality of the produced outcomes [1]. Data validation and cleaning are often employed to improve data quality [2]. One important practice in data cleaning is entity resolution, which is the task of identifying and matching all records that refer to the same real-world entity. An entity could be a person, a product, a business, or any other real-world object.

Entity resolution (ER) is challenging because databases usually do not contain unique entity identifiers. In this case identifying attribute values (such as first

---

* This research was funded by the Australian Research Council (ARC), Veda, and Funnelback Pty. Ltd., under Linkage Project LP100200079.

H. Wang and M.A. Sharaf (Eds.): ADC 2014, LNCS 8506, pp. 1–12, 2014.

names, surnames, or addresses) need to be used for the matching process. However, such attribute values are often of low quality, as they can be incomplete,
contain errors, or change over time [1]. Therefore, approximate string matching
techniques are generally required.

As services in both the private and public sectors move online, organizations
increasingly require to perform real-time ER (with sub-second response times)
on query records that need to be matched with existing entity databases [3].
These databases are often not static, but rather dynamic as queries generally
result in a record being modified, added, or even removed.

Most current ER techniques are, however, batch algorithms only suitable for
static databases. They compare and resolve all records in one or more database(s)
rather than resolving those relating to a single query record. There is a need for
new techniques that support ER for large dynamic databases that can resolve
streams of query records in real-time. A major aspect of achieving this goal
is to develop novel indexing techniques that allow dynamic updates and facilitate real-time matching by generating a small number of high-quality candidate
records.

Contributions: First, we propose a braided AVL tree [4] based technique
that facilitates dynamic indexing based on the sorted neighborhood method [5].
The proposed index can be dynamically updated with query records resulting
in up-to-date candidate sets in highly dynamic environments. Second, we investigate using fixed and adaptive window sizes when retrieving candidates from
neighboring tree nodes. Third, we improve matching times for query records
by proposing a pre-calculation of attribute value similarities between neighboring tree nodes, resulting in a significant reduction in matching times. Finally,
we experimentally evaluate our approaches using two real data sets with several
million records with personal details. These experiments show that our approach
significantly outperforms previous real-time matching techniques for ER [6,7].

2   Related Work

General similarity search approaches involve finding similar entities from unstructured databases (such as emails, news articles, or scientific publications)
based on a collection of relevant features that are represented as points in high-
dimensional attribute spaces [8,9]. However, such approaches are less suited for
structured databases that contain well defined attribute with short values, such
as personal names, addresses, or dates of birth. Records in structured databases
can be matched using SQL join statements if unique entity identifiers, such as
Medicare or social security numbers, are available. However, such identifiers are
often not available, and therefore ER approaches need to be employed [1].

The ER process encompasses several steps [1], including data pre-processing,
indexing or blocking, record pair comparison, record pair classification (into
matches and non-matches), and evaluation with regard to matching accuracy
and completeness. This paper is mostly concerned with the indexing step.

Standard blocking and the sorted neighborhood method (SNM) indexing techniques are commonly used in the ER process. Standard blocking [10] is based on inserting records into blocks according to a *blocking key* criteria and only comparing records that are in the same block. The SNM [5] arranges all records in the database(s) to be matched into a sorted array using a *sorting key* criteria. Then a fixed-size window is moved over the sorted records, comparing only those records that are within the sliding window at any one time. Both blocking and sorting keys are usually based on one or a concatenation of attribute values.

Various other indexing techniques have been used for ER, such as q-gram indexing, suffix array indexing, canopy clustering, and mapping-based indexing [1]. However, all these techniques are aimed at offline batch processing of static databases and are limited to indexing of static data. This means that once an index is created it is difficult to change if new records need to be added, or when the values in existing records are changing.

Only limited research has so far concentrated on real-time ER, where a stream of query records arrive that need to be resolved in sub-seconds against the records in a database, or on ER for dynamic databases. A first approach to query-time ER was based on a collective clustering technique [11]. The authors stated that the average query time was 31.28 sec on a database with around 800,000 records; this approach is therefore not suitable for real-time ER. Ioannou et al. [12] proposed an approach based on using links between the entities in a probabilistic database to resolve these entities. On a database of around 50,000 records the approach was reported to have an average query time of 70 msec. This approach works with dynamic databases and can be used for real-time ER.

Christen et al. [6] proposed a similarity-aware indexing technique where similarities between attribute values are pre-calculated when the index is built. An average query time of 10 msec was reported by the authors on a database with several million records. Although this index facilitates real-time ER it is only applicable for static databases. More recently, Ramadan et al. [7] extended this similarity-aware index to work with dynamic data. The authors stated that the growing size of the index is not affecting the average record insertion time which was around 0.1 msec on the same database as used in [6], and the average query time which was about 10 msec. While this dynamic similarity-aware indexing technique is based on standard blocking described before, we propose a novel dynamic real-time indexing approach based on the SNM.

3 Dynamic Sorted Neighborhood Indexing

The original SNM uses a static array data structure to store sorting key values (SKV) of all records in the database(s). Our proposed dynamic sorted neighborhood index (DySNI), on the other hand, uses a braided AVL tree, which is a data structure that combines the properties of both a height balanced binary tree and a double-linked list [4]. Each node in the braided AVL tree has a link to its predecessor and successor nodes according to an alphabetical sorting of the key values in the nodes. Figure 1 illustrates a braided tree for the small example data set shown in the same figure.

ID	FName	SName	Postcode
r1	percy	smith	10007
r2	paul	smith	02120
r3	robin	stevens	80202
r4	pedro	smith	90005
r5	abby	bond	10001
r6	sally	taylor	90002
r7	peter	smith	90012
r8	sally	taylor	98168
r9	pedro	smith	02121
r10	peter	smith	90002

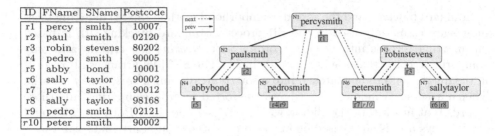

Fig. 1. The table on the left shows a small example data set. Record $r10$ is assumed to be a query record. The figure on the right represents the proposed DySNI built for records from the table. The SKV used is a concatenation of FName and SName values.

Because the key values in the tree are the SKVs of records, the identifiers of all records that have the same SKV will be appended to the corresponding tree node as a list (as shown in Figure 1). Assuming there are k different SKVs (nodes) in a tree, and n records in the database to be indexed (with $k < n$, potentially even $k \ll n$), searching for a SKV will be reduced from $O(log(n))$ to $O(log(k))$ compared to the array based SNM. The DySNI approach contains two phases, a *build phase* and a *query phase*.

Build Phase: During the build phase, records are loaded from a database, their SKVs are generated and inserted into the tree data structure, with each unique SKV becoming a node in the tree. Each node in the tree has a link to its alphabetically sorted predecessor node ('prev'), a link to its successor node ('next'), and a list of the identifiers of all records in the database that have that node's key value as their SKV. If the SKV of an inserted record is new, a new node will be created in the tree for this SKV. On the other hand, if the SKV for a certain record already exists in the tree as a node key, then only the identifier of this record needs to be added to the corresponding list.

After having loaded and indexed all records in a database, the index is ready for receiving and resolving query records. The complete records with full attribute values are also indexed into an inverted index or disk-based database table **D**, where the actual attribute values of records can be retrieved during the record comparison step, which is part of the query matching process.

Query Phase: In the query phase (shown in Algorithm 1), a query record **q** is to be matched against the built index in real-time. We assume that query records are added to the DySNI upon arrival. When a query record arrives, the first step is to generate the SKV for the record (line 1) and a new unique record identifier **q**.id is assigned to it. This SKV and record identifier are then inserted into the tree in the same way as records were inserted during the build phase (lines 3-7), and the query record is also added into **D**.

The window of neighboring nodes can now be generated (line 8). All record identifiers that are stored in the nodes within the window are added to the candidate record set **C**. The query record **q** will be compared with the records in **C** using similarity comparison functions, such as approximate string comparisons [1],

Algorithm 1: *DySNI − Query*	Algorithm 2: *SimDySNI − Query*
Input:	Input:
- Query record: **q**	- Query record: **q**
- Similarity functions: **S**	- Similarity functions: **S**
- Sorting key attributes: **SK**	- Sorting key attributes: **SK**
- Window size: **w**	- Window size: **w**
- Database table with complete records: **D**	- Database table with complete records: **D**
Output:	Output:
- Ranked list of matches: **M**	- Ranked list of matches: **M**
1: **skv** = $GenerateKey($**SK**, **q**$)$	1: **skv** = $GenerateKey($**SK**, **q**$)$
2: **nd** = $FindTreeNode($**skv**$)$	2: **nd** = $FindTreeNode($**skv**$)$
3: **D**[**q**.*id*] = **q**	3: **D**[**q**.*id*] = **q**
4: **if nd** == NULL **then**	4: **if nd** == NULL **then**
5: **nd** = $CreateNode($**skv**, **q**.*id*$)$	5: **nd** = $CreateNode($**skv**, **q**.*id*$)$
6: **else**	6: $PreCalcNodeSimilarities($**nd**, **w**$)$
7: Append **q**.*id* to **nd**.*id_list*	7: $UpdateSimNextNodes($**nd**, **w**$)$
8: **C** = $GenerateWin($**nd**, **w**$)$	8: $UpdateSimPreviousNodes($**nd**, **w**$)$
9: **M** = $CompareRecords($**C**, **S**, **D**, **q**$)$	9: **else**
10: Sort **M** according to similarities	10: Append **q**.*id* to **nd**.*id_list*
	11: **C** = **nd**.*nd_id_list* ∪ **nd**.$GetPrevIdList()$
	∪ **nd**.$GetNxtIdList()$
	12: **M** = $ComparePreCalcRecords($**C**, **S**, **D**, **q**$)$
	13: Sort **M** according to similarities

appropriate to the content of each attribute (line 9). The actual attribute values are retrieved from the index or database table **D**. The compared candidate records are returned in the list **M** sorted according to their overall similarities with the query record (line 10).

3.1   Generating the Window of Neighboring Nodes

To generate the window of neighboring nodes we investigated three approaches: the first is based on a static window with a fixed pre-defined size, while the second and third approaches are based on adaptive window sizes that vary based on the characteristics of tree nodes.

Fixed Window Size. The original SNM is based on using a fixed size window w that corresponds to the number of candidate records that falls inside the window at any one time. As our DySNI approach is a tree-based index, and because all records that have the same SKV are inserted into one node, we set w as the number of neighboring tree nodes in one direction (previous and next). The following example demonstrates the query phase of DySNI.

The index tree shown in Figure 1 is built on all records from the table in the same figure. Assuming query record $r10$ has just been inserted into the tree in node $N6$ with key value 'petersmith', and assuming a fixed window size $w = 1$ in each direction, the record identifiers from only one neighboring node of $N6$ are retrieved. The previous node is $N1$ with key value 'percysmith' and record identifier $r1$, while the following (next) node is $N3$ with key value 'robinstevens' and record identifier $r3$. The final set of candidate records for query record $r10$ using this fixed window size approach is the set $\mathbf{C} = \{r1, r3, r7\}$. Note that $r7$ is included as it is located in the same tree node as the query record, and so it also needs to be compared with the query record.

As can be seen from this example, a fixed size window can lead to both unnecessary comparisons with records in nodes that unlikely will have a high enough similarity to be matching with a given query record (like $r3$ from node $N3$), as well as missed potential true matches that are outside the window (such as the records attached to node $N5$ with key value 'pedrosmith').

Adaptive Window Size. The aim of using an adaptive window size is to limit the number of comparisons between the query and candidate records to only those records that likely correspond to true matches. This issue was addressed for static SNM by two approaches that adjust the window size according to the characteristics of the SKVs. The first approach expands the window based on the similarities between SKVs [13], while the second approach expand the window based on the number of classified matches within the window [14]. Here we propose two adaptive approaches that can work with our DySNI.

Similarity-Based Adaptive Window: This approach is based on [13]. We adaptively expand a window on each side of a query record's tree node individually based on the following steps. We start from the node that contains the query record. Then we expand the window in each direction based on the similarity between SKVs, until the similarity between the query's SKV and the SKV of nodes in a direction falls below a given minimum similarity threshold θ ($0 \leq \theta \leq 1$, with $\theta = 1$ being an exact match). The aim of this approach is to exclude tree nodes that likely contain records that are dissimilar to the records in the query record's tree node.

The following example explains the approach based on Figure 1 with $\theta = 0.6$. After inserting $r10$ into the index, generating the window of candidate records starts from the query node $N6$. To expand the window forwards (next), we compare the SKV of node $N6$ with the SKV of its next neighbor $N3$ with SKV 'robinstevens' using the edit distance approximate string comparison function [1]. This gives us a low value of $sim('petersmith', 'robinstevens') = 0.25$. Because $0.25 < \theta$ the window does not expand forward and the node $N3$ and its record identifier list ($r3$) will not be included into the set of candidate records **C**.

The same process will take place in the node's backwards (previous) direction. We get the SKV of the previous node $N1$ and compare it to the SKV of the query node, which leads to $sim('petersmith', 'percysmith') = 0.7$; so $N1$ and its record identifier $r1$ is added to the list of candidate records. The comparison process continues in this direction until we reach a similarity that is less than θ. This occurs at node ($N4$) where $sim('abbybond', 'petersmith') < \theta$. This means all records in the nodes $N5$ and $N2$ are included into **C**. The final set of candidate records from both sides will be $\mathbf{C} = \{r1, r2, r4, r7, r9\}$.

Candidates-Based Adaptive Window: This approach aims at getting a certain maximum number of nearest candidate records as can be processed within a certain period of time. In a real-time environment this approach allows for a controlled number of candidate records to be returned for detailed comparisons. A threshold β, which is a minimum total number of candidate records to be returned, is used to stop window expansion regardless of the similarities between

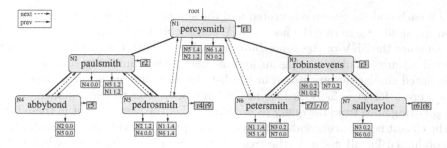

Fig. 2. Similarity-based DySNI built for records from the table in Figure 1. The same SKV is used, and the window size is set as $w = 2$.

SKVs. The initial candidate record set **C** is the records located in the query record's node. Then a decision on whether to expand the window on both sides or not is made based on the following.

If the count of records at the query record's node is greater or equal to the minimum candidate threshold $|C| >= \beta$, then no expansion is needed, and only records located at the query node are included in **C**. On the other hand, if $|C| < \beta$, then the window expands on both sides of the initial node individually until $|C| >= \beta$. The remaining number of records needed for the total candidate records to reach β is calculated as $r = \beta - |C|$. A new expansion threshold is set to $\lceil r/2 \rceil$ for each side of the query node, and the window on each side will continue expanding as long as the total number of candidate records from that side is smaller than or equal to $\lceil r/2 \rceil$.

3.2 Similarity-Based Dynamic Sorted Neighborhood Indexing

This similarity-based dynamic sorted neighborhood index (SimDySNI) is based on the knowledge that all records allocated to the same tree node have the same value in their SKV. As several attributes are generally used as SKV, we can pre-calculate the similarities between these attribute values in neighboring tree nodes at the end of the build phase of the index. To reduce the time required for the calculation of similarities between records in the query phase, the pre-calculated similarities are used when a query record is compared with a set of candidate records.

In the SimDySNI, additional to the basic braided AVL tree structure of the DySNI, each node will have two extra lists attached, as is illustrated in Figure 2. These lists contain the pre-calculated similarities for a fixed window size w. In the following we describe how a SimDySNI tree is built and how it is queried using a fixed size window. Investigating the SimDySNI for the two adaptive approaches is left for future work.

Build Phase: After the build phase is completed (as previously explained for DySNI), a similarity calculation phase is conducted where the pre-calculated similarities are added into the built DySNI tree. In this similarity calculation phase each node in the tree is visited and the similarities between the attributes that are used to generate the node's SKV and the attribute values that are used to generate the SKV of the w neighboring nodes (in both the previous and next direction) are calculated using an approximate string similarity function. The calculated similarities are stored in two lists for each tree node, as is illustrated in Figure 2 for $w = 2$. Both lists are ordered according to the distance of the neighboring node from the query record's node (i.e. the first element in these lists is the closest neighboring node, and so on). The process of calculating similarities is conducted for all nodes in the tree. At the end of this phase the SimDySNI can be queried as we describe next.

Query Phase: This phase (illustrated in Algorithm 2) differs from the query phase in the DySNI in that the similarities between the query and candidate records only have to be calculated for those attributes that are not used in the sorting key (**SK**). A query record is first inserted into the index data structure in the same way as was done with the DySNI (lines 3-10). If the query record has a SKV that is new (which required creating a new node for this SKV), then we need to calculate the similarities for its w next and previous neighboring tree nodes (as described for the similarity calculation phase above) (line 6). We also need to update the similarity lists for the w previous and next tree nodes of the newly inserted tree node (lines 7-8). This step ensures that the pre-calculated similarities are up-to-date at any time.

To generate candidate records for a query record **q**, we retrieve all records that are stored in the tree nodes from the next and previous lists of the node that holds the query record (line 11). To calculate the overall similarities between the query record and candidate records we retrieve the pre-calculated similarities from the previous and next similarity lists, retrieve the corresponding records from the record identifier lists of these tree nodes, and then only calculate the similarities of those attributes that are not used in the sorting key. Therefore, the more attributes are used in a sorting key the more similarities can be pre-calculated, but at the cost of a larger tree (as likely more unique SKVs will be generated). In our experimental evaluation we investigate how different sorting keys influence both the amount of memory required as well as the reduction in query matching time that can be achieved.

The following example describes the query phase on the small set of records from the table in Figure 1 and the SimDySNI tree shown in Figure 2 assuming $w = 2$. The candidate records for query record $r10$ are the records from the nodes stored in the two similarity lists of node $N6$. These are the records from nodes $N1$ and $N5$ (previous), and $N3$ and $N7$ (next). The total set of candidate records for query $r10$ will therefore be $\mathbf{C} = \{r1, r3, r4, r6, r7, r8, r9\}$. To compare query record $r10$ with these candidate records we first retrieve the actual records from $\mathbf{D}$, and for each candidate record we retrieve the pre-calculated similarity from the SimDySNI index. For example, the pre-calculated similarity between query record $r10$ and candidate record $r1$ in $N1$ is 1.4 as retrieved from the previous list of node $N6$. This similarity corresponds to the pre-calculated edit-distance similarities of the FName and SName attributes (each is between

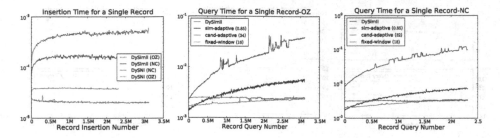

Fig. 3. Average insertion and query time of the proposed approaches compared to DySimII using the full OZ and NC data sets [7]. The y axis shows times in seconds.

0 and 1). To get the total similarity between $r10$ and $r1$ we then only have to calculate the similarities for the Postcode attribute on the attribute values of these records.

4 Data Sets and Experimental Evaluation

To evaluate different aspects of our proposed DySNI we used two large data sets. The first is a real voter registration data set from the US state of North Carolina (named 'NC') [15]. We have downloaded this data set every two months since October 2011 to build a temporal data set that contains the names, addresses, and ages of over 2.5 million voters, as well as their voter registration numbers. This data set contains realistic temporal information about a large number of people. We identified 111,403 individuals with two records, 2,408 with three, and 39 with four records in this data set.

A second data set is based on an Australian telephone directory (named 'OZ') which contains nearly 3.5 million records. This data set does not contain any real duplicate entries. We therefore created duplicate records by modifying the values in selected attributes using edits based on keying mistakes [16].

We implemented all evaluated approaches using Python (version 2.7.3) and ran all experiments on a server with 128 GBytes of main memory and two 6-core Intel Xeon CPUs running at 2.4 GHz.

4.1 Results and Discussion

In our first set of experiments, we evaluated whether the proposed DySNI scales to large and dynamic databases while facilitating real-time ER, where fast query matching response times are required. We measured the average time required to insert a single record into an index data structure, and the average query time required to resolve a single query record across the growing size of the index tree structure. These experiments were conducted on both the full OZ and NC data sets. The DySNI with the fixed-size window and the two adaptive window approaches were compared to the dynamic indexing technique DySimII which is based on standard blocking. [7]. The window sizes and thresholds selected

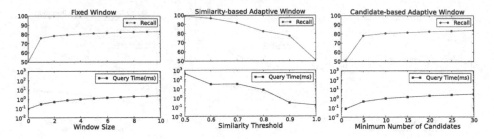

Fig. 4. Recall and average query time for the proposed window generating approaches. These plots are generated using 351,876 records (10%) from the OZ data set.

(Figure 3) for running the experiments assures having the same average number of comparisons for all window approaches of the proposed DySNI.

As can be seen from Figure 3, the DySNI approaches (both fixed and adaptive) significantly outperform the earlier DySimII by up-to one magnitude faster insertion times, and nearly two magnitudes faster query times. The average insertion time is also not affected by the growing size of the index data structure, while the query time only increases slightly as the index becomes larger. The results show that the different variations of the proposed DySNI approach achieve very fast query times that range between 0.02 and 0.3 msec per query record.

In our second set of experiments we evaluated the three proposed window approaches using different threshold values. The aim of this set of experiments is to investigate the effect of using different thresholds on recall (calculated as the ratio between identified matches and true matches over all query records), and on average query time. These experiments were conducted using 351,876 records from the OZ data set (i.e. 10% of the full data set).

As can be seen in Figure 4, the similarity-based adaptive DySNI achieves higher recall with slower query times compared to the fixed window and candidate-based DySNI. We also notice that a lower similarity threshold gives better recall but requires longer query times, because more tree nodes are included in the window and thus more candidate records are generated. However, although the average query time achieved by the similarity-based DySNI is larger than for the other two approaches, the slowest time achieved is less than 0.4 sec.

For both fixed-window and candidate-based approaches larger window sizes and larger values of the minimum number of candidate records gives better recall results at the cost of unnecessary comparisons that increases the query time. Since expanding the window is not dependent upon the characteristics of the records within tree nodes, the expansion process in both approaches could stop even if there are nearby matches outside the window.

In the last sets of experiments we investigated how the SimDySNI is able to reduce query time. The attributes used for comparing a query record with candidate records were first name, surname, city, and postcode from the OZ data set. We measured the average time needed to compare a query record with a single candidate record for DySNI, and for the SimDySNI approach with different

Data set	Num nodes	MB DySNI	MB SimDySNI
SKV: FName			
OZ	167,558	101	135
NC	120,632	74	98
SKV: SName			
OZ	475,934	238	333
NC	194,090	106	144
SKV: Postcode			
OZ	5,598	30	34
NC	508	21	21
SKV: SName+FName			
OZ	2,589,462	1,192	1,705
NC	1,634,650	754	1,079
SKV: SName+FName+Postcode			
OZ	3,335,313	1,543	2,205
NC	2,237,069	1,130	1,614

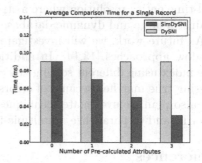

Fig. 5. The table shows the number of nodes in the generated trees, as well as the memory usage in MBytes required for DySNI and SimDySNI using both the full OZ and NC databases. The plot shows the average comparison time required for DySNI and SimDySNI approaches using 10% of records from the OZ data set.

numbers of attributes used in the sorting key. We again ran experiments on 10% of the OZ data set with 1, 2 and 3 attributes used as sorting keys for different possible combinations of compared attributes. The average query times over these combinations are shown in Figure 5. The results show that the SimDySNI can significantly reduce the time required to compare a query record with a single candidate record. This improvement in time is almost linear with the number of attributes used in a sorting key. The result show that for a one-attribute sorting key the query time reduction is around 20%, for a two-attribute sorting key it is around 40%, and for a three-attributes sorting key it can be up-to 70%.

5   Conclusions

We have presented a tree-based sorted neighborhood indexing technique which is suitable for real-time ER on large dynamic databases. We investigated using both fixed and adaptive window sizes to generate the candidate records. The proposed techniques were evaluated using two large data sets.

This evaluation showed that the growing size of the proposed DySNI has no significant effect on the average insertion and query times. The similarity-based adaptive DySNI showed better results than the other two approaches because the expansion decision in this approach depends on the similarities between SKVs. The fixed window and the candidate-based adaptive approaches showed better timing results, but at the cost of achieving lower recall values. This is because they only consider the number of comparisons to be conducted rather than the similarity between values. Also, our results illustrate that using the similarity-based SimDySNI reduces the average comparison time between 20% to 70% (based on the number of attributes used to generate SKV), while it increases the memory footprint by 13% to 40% for different SKVs. However, the memory needed for the SimDySNI is still small and ranges between 34 MB and 2,205 MB for the various SKVs used to build the tree on over 3 million records. All discussed results confirm that the proposed DySNI is well suited for use with real-time entity resolution where a stream of query records needs to be resolved against a large and dynamic database.

As future work, we will investigate combining SimDySNI with the adaptive window approaches [13,14]. In addition, we will investigate building a multi-tree index using different sorting key combinations to overcome the drawback of missing true matches because of errors and variations at the beginning of SKVs. We also plan to investigate techniques to learn optimal tree selection for query records, and to parallelize a multiple-tree index to improve performance.

References

1. Christen, P.: Data Matching: Concepts and Techniques for Record Linkage, Entity Resolution, and Duplicate Detection. Springer (2012)
2. Herzog, T., Scheuren, F., Winkler, W.: Data quality and record linkage techniques. Springer (2007)
3. Dong, X.L., Srivastava, D.: Big data integration. In: IEEE ICDE, Brisbane, AU, pp. 1245–1248 (2013)
4. Rice, S.V.: Braided AVL trees for efficient event sets and ranked sets in the SIM-SCRIPT III simulation programming language. In: Western MultiConference on Computer Simulation, San Diego, pp. 150–155 (2007)
5. Hernandez, M.A., Stolfo, S.J.: The merge/purge problem for large databases. In: ACM SIGMOD, San Jose, pp. 127–138 (1995)
6. Christen, P., Gayler, R., Hawking, D.: Similarity-aware indexing for real-time entity resolution. In: ACM CIKM, Hong Kong (2009)
7. Ramadan, B., Christen, P., Liang, H., Gayler, R.W., Hawking, D.: Dynamic similarity-aware inverted indexing for real-time entity resolution. In: Li, J., Cao, L., Wang, C., Tan, K.C., Liu, B., Pei, J., Tseng, V.S. (eds.) PAKDD 2013 Workshops. LNCS (LNAI), vol. 7867, pp. 47–58. Springer, Heidelberg (2013)
8. Gionis, A., Indyk, P., Motwani, R.: Similarity search in high dimensions via hashing. In: VLDB, Edinburgh, Scotland, pp. 518–529 (1999)
9. Zhang, Z., Jiang, J., Liu, X., Lau, R., Wang, H., Zhang, R.: A real time hybrid pattern matching scheme for stock time series. In: Australin Database Conference, pp. 161–170. Australian Computer Society, Inc., Brisbane (2010)
10. Fellegi, I., Sunter, A.: A theory for record linkage. Journal of the American Statistical Association 64(328), 1183–1210 (1969)
11. Bhattacharya, I., Getoor, L.: Query-time entity resolution. Journal of Artificial Intelligence Research 30, 621–657 (2007)
12. Ioannou, E., Nejdl, W., Niederée, C., Velegrakis, Y.: On-the-fly entity-aware query processing in the presence of linkage. VLDB Endowment 3(1) (2010)
13. Yan, S., Lee, D., Kan, M.Y., Giles, L.C.: Adaptive sorted neighborhood methods for efficient record linkage. In: ACM/IEEE-CS Joint Conference on Digital Libraries, Vancouver, Canada, pp. 185–194 (2007)
14. Draisbach, U., Naumann, F., Szott, S., Wonneberg, O.: Adaptive windows for duplicate detection. In: IEEE ICDE, Washington, DC, pp. 1073–1083 (2012)
15. Christen, P.: Preparation of a real voter data set for record linkage and duplicate detection research. Technical report. Australian National University (2013)
16. Christen, P., Pudjijono, A.: Accurate synthetic generation of realistic personal information. In: Theeramunkong, T., Kijsirikul, B., Cercone, N., Ho, T.-B. (eds.) PAKDD 2009. LNCS (LNAI), vol. 5476, pp. 507–514. Springer, Heidelberg (2009)

Efficient Aggregate Farthest Neighbour Query Processing on Road Networks

Haozhou Wang, Kai Zheng, Han Su, Jiping Wang, Shazia Sadiq,
and Xiaofang Zhou

School of ITEE, The University of Queensland,
St. Lucia, Brisbane, QLD 4072, Australia
{h.wang16,kevinz,h.su1,j.wnag28,shazia,zxf}@uq.edu.au

Abstract. This paper addresses the problem of searching the k aggregate farthest neighbours (AkFN query in short) on road networks. Given a query point set, AkFN is aimed at finding the top-k points from a dataset with the largest aggregate network distance. The challenge of the AkFN query on the road network is how to reduce the number of network distance evaluation which is an expensive operation. In our work, we propose a three-phase solution, including clustering points in dataset, network distance bound pre-computing and searching. By organizing the objects into compact clusters and pre-calculating the network distance bound from clusters to a set of reference points, we can effectively prune a large fraction of clusters without probing each individual point inside. Finally, we demonstrate the efficiency of our proposed approaches by extensive experiments on a real Point- of-Interest (POI) dataset.

1 Introduction

As one of the most important types of spatial query, efficient nearest neighbour query processing has been investigated extensively[1,2,3]. This type of query to find k nearest neighbors (kNN) form a given point becomes a fundamental operation in spatial databases, leading to a number of variations. As much as the interest in finding kNN objects, there are a large number of real-life applications which are interested in finding farthest neighbours (FN) [4]. Another important type of kNN variations is the so-called aggregate nearest neighbour (ANN) query [5,2]. The difference between ANN query and NN query is that an ANN query takes multiple query points into account and returns a point from the dataset that minimizes the aggregated distance from the point to all given query points using a user-specified aggregate function (e.g. *max*, *sum*). An example of ANN query with *sum* function is for a number of people to find a meeting place that can minimize their total traveling distance.

In an analogous way, aggregate farthest neighbor (AFN) query can also be defined as an extension to the FN query: for a given set of query points Q and a user-specified aggregate function, find a point p from a data set P such that the aggregate distance from p to all the points in Q is the *largest*. AkFN can be defined as a general case of AFN to find k such points which have larger

H. Wang and M.A. Sharaf (Eds.): ADC 2014, LNCS 8506, pp. 13–25, 2014.
© Springer International Publishing Switzerland 2014

aggregate distances than any other points in P. To illustrate its usefulness, consider a business franchise planning to open a new store. In order to reduce the mutual influences between the new and existing branches to maximise the overall profit, it is desirable for the location of the new store to be far away from all existing stores. By including the locations of all existing stores in Q, all the available locations from a real estate database as P and max as the aggregate function, an AkFN query can find the best candidate locations to choose from.

Despite of its importance for many applications, AkFN query has not been well studied. Just like ANN query processing is a non-trivial extension to NN query processing, AkFN query processing is quite different form kFN query processing and demands new processing strategies. To the best of our knowledge, there exists only one piece of work on AkFN query processing [6]; it, however, considers a simpler case to use Euclidean distance (i.e., in a free space). In this paper, we will investigate AkFN in the context of road networks. The motivation for us to consider road networks is that, in most real applications the movement of people and vehicles is constrained by a underlying road network. Albeit more complex in distance calculation, it is more reasonable and accurate to use network distances rather than Euclidean distances. This is because, in reality, road network contains some properties such as bridges and one way street, which makes the distance shortest path between two points in road network is longer than its Euclidean distance. Therefore, the incorporation of road networks can raise serious efficiency issues for processing AkFN query. The reasons can be two-fold. First, there is still no effective way to index a large number of objects in a road network. Classical hierarchical spatial access methods (e.g., R-tree [7] based) cannot work since they are designed for Euclidean space. Second, network distance evaluation is much more expensive than Euclidean distance evaluation since it involves online shortest path computation.

Our paper aims to propose efficient solutions for answering the AkFN query in road networks. More specifically, we firstly organize the objects in the whole dataset into clusters by apply a network-based hierarchical clustering method. Then we define a set of reference points across the entire space and pre-compute the maximum and minimum network distances between each pair of cluster and reference point. Lastly, we design an efficient search algorithm to spot the most promising clusters that may contain the results and prune the rest of them by leveraging the pre-computed information. It is worth noting that, since the clusters are hierarchical, we can achieve good trade-off between the number of clusters and pruning effect. In summary, we make the following major contributions in this paper: 1) We are the first to investigate the aggregate farthest neighbour query in the context of road networks. 2) We propose efficient solutions for processing the AkFN query by pre-computing and pruning at cluster level. 3) We conduct extensive experiments based on real POI dataset to verify the efficiency of our proposals.

The rest of this paper is organized as follows. In Section 2 we will introduce necessary preliminaries and formally define the AkFN query. We detail our proposed query processing algorithms in Section 3. Section 4 presents the experimental results for validating the efficiency of our algorithms, followed by a brief literature review on related work in Section 5. We conclude the paper in Section 6.

2 Problem Definition

In this section we will introduce all necessary preliminary concepts and formulate the AkFN query. We summarize the major symbols and notations used throughout the paper in Table 1 for convenience of reference.

Table 1. Table of Notations

Notation	Definition
(v_i, v_j)	A road segment with two road segment nodes v_i and v_j
p_i	A point in the points dataset P
q_i	A query point in the query points set Q
$d_n(p_i, p_j)$	The network distance between p_i and p_j
$d_{agg}(p_i, Q)$	The aggregate distance from a point p_i to Q
r_i	A reference point in the reference points set R
$dr(p_i, q_i)$	The shortest path distance from p_i to q_i via q_i's reference point
C_i	A sub-cluster in hierarchical cluster structure C
$dc(C_i, r_i)$	The maximum network distance from C_i to r_i
$dr(C_i, q_i)$	The network distance from C_i to q_i via q_i's reference point
$dr_{agg}(p_i, Q)$	The aggregate distance from p_i to Q via relate reference points
$dr_{agg}(C_i, Q)$	The aggregate distance from C_i to Q via relate reference points

Definition 1 (Road Network and Network Distance). *A road network G is modeled as a weighted indirect graph $G = (V, E)$, where V is a set of road intersection , and E is a set of road segment. The network distance d_n between p_a and p_b, where p_a and p_b are two points on G, is calculated as the sum of the distance of the road segments along the shortest path between p_a and p_b.*

Notice that we use the term "network distance" and "shortest path distance" interchangeably.

Definition 2 (Aggregate Network Distance). *Given a point p, a query point set Q and road network G, the aggregate network distance between p and Q is $d_{agg}(p, Q) = f_{q \in Q} \, d_n(p, q)$, where f is a pre-defined aggregate function (e.g., sum, mean, min, max, etc).*

In this paper we only consider two types of aggregate functions, namely *sum* and *min*, as they are most applicable in a farthest neighbor query. Now we are in a position to formally define the query.

Definition 3 (AkFN query). *Given a dataset P and a query point set Q, the aggregate k farthest neighbor (AkFN) query retrieves a set S of k points from P that have the largest aggregate network distance with Q, i.e.,*

$$d_{agg}(p, Q) \geq d_{agg}(p', Q), \quad \forall p \in S, \forall p' \in P - S$$

Consider Figure 1 as an example, where q_1, q_2 is the query set Q and $p_1, p_2 \ldots p_8$ is the candidate dataset P, that we want to find a point from P, such that its minimum distance to Q is maximized. This is a special case of the AkFN query with $k = 1$ and min aggregate function. By enumerating the locations in P and simple calculation, it is easy to get that p_7 is the best location suiting for the request.

3   Query Processing Algorithm

The most straightforward approach to answer an AkFN query is to exhaustively search all the points in dataset, calculate the aggregate distances from each point to the query point set, and finally obtain the top k results. However, this method, called *exhaustive search* algorithm, has serious efficiency issues especially on road network, since the exhaustively search algorithm need to search the whole dataset P while P is usually very large (e.g., more than 100k points) in practice. Consequently, evaluating the aggregate network distance between all points in the dataset and the query "on-the-fly" can be extremely time consuming.

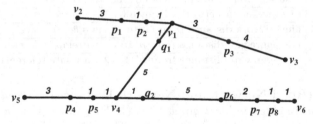

Fig. 1. Example of road network

To improve the efficiency of query processing, we propose an advanced approach with two carefully designed search algorithms, which leverage the power of hierarchical cluster and pre-computed network distance bound of each cluster to reference points to reduce the search space extensively. The query processing consist of three steps: clustering the points of the dataset, pre-computing the network distance bound and searching. The first step clusters the points of the dataset in hierarchical structure by applying the Linkage Hierarchical Clustering algorithm [8], a network-distance-based clustering method. In the second step, we uniformly define a set of *reference points*, which are mapped to road segment nodes, and pre-compute the network distance bounds between each pair of the clusters, generated in the first step, and reference points. This information can be saved for further use since the numbers of both clusters and reference points are relatively small comparing to the original dataset. The third step is searching the hierarchical structure with two different ways by search algorithms. In the rest of this section, we will describe each step in detail.

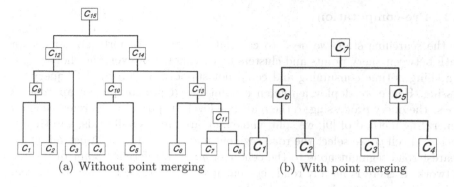

(a) Without point merging   (b) With point merging

Fig. 2. Example of hierarchical cluster

3.1 Linkage Hierarchical Clustering

Using Euclidean-distance-based index (i.e. R-tree [7]) to prune unnecessary points is unsafe for AkFN query, since two points are close on Euclidean space do not mean they are close on road network. For example, two points are located in each side of a river and a bridge is located far away from these two points, absolutely, their network distance is much longer than their Euclidean distance. Obviously, in order to improve the performance of AkFN, a road-network-based point organizing structure is needed.

Thus, we adopt the Linkage Hierarchical Clustering algorithm [8], a well-known road-network-based clustering method, in our search approaches. The motivation for us to use hierarchical clustering is the trade-off between the number of clusters and the searching performance. Though a smaller number of bigger clusters can reduce the overhead cost incurred by storing and processing these clusters, the pruning effect will also be harmed since the distance bound is too loose to be useful. By organizing the objects into clusters with different levels and sizes, we can control the level from which the search starts and thus achieve a reasonable balance between the cluster number and pruning effect. Fig. 1 gives an example of a point dataset $P = p_1, p_2, \cdots, p_8$ on road network and its hierarchical cluster structure is shown in Fig. 2(a), which each cluster in the bottom level is the point itself in P.

However, consider each point of dataset P as a cluster at initiating time of the above algorithm can result in too many bottom level clusters when the size of P is big, which may cause memory overflow in pre-computing step and inefficient query processing in searching step. In order to reduce memory consumption and speed up the query processing, we merge points of P before applying clustering, if the road network distances between such points are less than a threshold ϵ. Fig. 2(b) shows an example of the hierarchical cluster structure with points merging for previous example (Fig. 1). With $\epsilon = 3$, p_1, p_2 are merged as C_3; p_3 cannot be merged with other points and considered as C_2. Meanwhile, p_4, p_5 and p_6, p_7, p_8 are merged as C_3 and C_4 respectively.

3.2   Pre-computation

In the searching step, we need to calculate distance of shortest road network path between query points and clusters many times. However, the shortest path searching is time consuming and could not support the "on-the-fly" query processing. Hence we deploy a reference point based pre-computing approach to boost the query processing to the real-time level. Inspired by the reference point generating method of [9], we split whole map area into small grids, and then for each grid cell gc we select the road segment node v, where v is in gc and v is the nearest road segment node to the center of gc, as a reference point of the road network. If there is no any road segment node in a grid cell, then no reference point will be selected in that grid cell.

The pre-computing is to calculate the maximum shortest path distance $dc(C, r) = \max_{p \in C}\{d_n(p, r)\}$ between each pair of cluster C of all the hierarchical clusters and point r of reference points set R. Therefore, the space cost of building reference list of pre-computed distance is $O(mn)$, where n is the number of clusters in the bottom level of the hierarchical structure, which is much less the total number of points in the road network; and m is the number of reference points. Meanwhile, the distance between reference points and upper level clusters can be calculated directly by previous information.

When a query point q is given, we find its nearest reference point r^*, that is $r^* = \arg\min_{r \in R}\{d_n(q, r)\}$. Then the $dr(C, q) = dc(C, r^*) + d_n(q, r^*)$ can be the upper bound of the shortest path distance $d_n(C, q)$ according to the triangle inequality. The upper bound can efficiently filter out some obviously impossible points during farthest neighbor search, which will demonstrate in the next section. Meanwhile, it is worthy to note that the more reference points mapped to road network, the more tight upper bound we can have, since the query point q could be much closer to reference point r and leads that $dr(C, q)$ is closer to $d_n(C, q)$. On the other hand, a lot of reference points means a heavy memory usage and may excess the total memory usage due to all reference points are loaded into memory during query processing. Hence, we limit the number of reference points, which suit for system memory or user request, to map in the road network, and we conduct a set of experiments to show the effect about number of reference points.

3.3   Search Algorithm

We propose two advanced search algorithms, namely Flat Search (FS) and Hierarchical Search (HS), which make use of the hierarchical clusters and pre-computed network distance bounds to improve the searching efficiency. FS only searches the bottom hierarchy of these clusters for the result; HS searches the whole hierarchy clusters.

Flat Search. We propose the FS algorithm by using the bottom hierarchical clusters to reduce the search space on road network. When given a query points set Q, we can quickly find the nearest reference point r_i of q_i ($q_i \in Q$) and the shortest path distance between q_i and r_i.

Firstly, FS initializes an empty list A with a fixed length $|k|$, where k is the number of top answers required by user. The elements of A are the result points so far, and sorted by aggregate distance $d_{agg}(p_c, Q)$. Meanwhile, we assign a parameter kth_so_far which is used to record the smallest $d_{agg}(p_c, Q)$in A to 0 . The value of kth_so_far can be a lower bound of the farthest neighbour query, that any point p' with the aggregate distance $d_{agg}(p_c, Q) < kth\_so\_far$ will be filtered out. Then, the FS starts at a random cluster C_i in bottom layer of hierarchical cluster, it calculates the aggregate distance from C_i to Q (denote as $dr_{agg}(C_i, Q)$). To calculate $dr_{agg}(C_i, Q)$, we firstly find the r_i, that r_i is the nearest reference point of q_i ($q_i \in Q$), and the shortest path distance $d_n(q_i, r_i)$ between q_i and r_i. Since the shortest path distances $dc(C_i, r_i)$ has already been pre-computed, FS computes the dr_{C_i, q_i}, which equals $dc(C_i, r_i) + d_n(q_i, r_i)$. If $dr_{agg}(C_i, Q)$ is larger than kth_so_far, it means that C_i may have a point which can be a result of the query. Otherwise, this cluster C_i should be pruned and FS selects the next cluster. Then, FS visits every point p_i in C_i, and computes the exact aggregate distance $d_{agg}(p_i, Q)$. If there is a $d_{agg}(p_i, Q)$ that is bigger than kth_so_far, FS insert this point in to A; and update kth_so_far. FS searches all of clusters in the bottom level of hierarchical cluster iteratively, and stops when all the bottom hierarchical clusters have been selected.

Hierarchical Search (HS). Flat search algorithm starts at a random cluster, which means it may have to scan all of the points in the worst case. In addition, calculating aggregate distances between of all clusters and query set are time-consuming, more important is that I/O cost of calculation aggregate distances for clusters and the query set Q is non-trivial and needs to be minimized. Hence, we improved our FS algorithm to implement on hierarchical clusters, which is called Hierarchical Search algorithm to minimize calculation cost of aggregate distances.

The key idea of HS is that HS maintains a priority queue to obtain the candidate set and adapts the best-first search. The elements of PQ are clusters C_i or points p_i and sorted by their $dr_{agg}(C_i, Q)$ or $dr_{agg}(p_i, Q)$ in descending order. In the first step, HS initializes a priority queue PQ. Then HS extracts the top layer cluster of hierarchical cluster; for each cluster C_i, HS calculates their $dr_{agg}(C_i, Q)$ and pushes them into priority queue PQ. During second phase, HS pops elements from PQ iteratively. For each element, HS compares its $dr_{agg}(C_i, Q)$ or $dr_{agg}(p_i, Q)$ with kth_so_far, if the $dr_{agg}(C_i, Q)$ or $dr_{agg}(p_i, Q)$ is larger than kth_so_far, this element is selected as a candidate element, therefore HS goes to next step. In this step, if this candidate element is a cluster that contains only one point (i.e., size of cluster is 1) or is a point, and then HS inspects this point p_i in this candidate cluster to calculate aggregate distance $d_{agg}(p_i, Q)$. If not, HS extracts this cluster to get its child clusters/points set, for each cluster C_i or point p_i in child clusters set, HS calculates $dr_{agg}(C_i, Q)$ or $dr_{agg}(p_i, Q)$ and push these elements to PQ with their aggregate distances as new candidate elements. After this, HS returns to the beginning of the second loop and continue popping the elements until $|A| = k$.

Given an example, at beginning, HS calculates cluster aggregate distance for top level clusters C_6, C_5 and push them to PQ, and elements in PQ are $\{(C_5, 8), (C_6, 4)\}$. Then first element $(C_5, 8)$ is popped, since kth_so_far is smaller than C_6's aggregate distance and the number of points in C_5 are larger than one. Then, HS gets C_5's children clusters, which is C_3 and C_4, and calculates their aggregate distance $dr_{agg}(C_3, Q)$ and $dr_{agg}(C_4, Q)$, push them into PQ. After that, the elements in PQ are $\{(C_4, 8), (C_6, 4), (C_3, 3)\}$. Similar with previous step, C_4 is popped, and elements in PQ are $\{(p_7, 9), (p_8, 8), (p_6, 6), (C_6, 4), (C_3, 3)\}$. Continually, point p_7 is popped, however p_7 is already a point, hence we calculate aggregate distance between p_7 and Q, and we get $d_{min}(p_7, Q) = 8$. After that, kth_so_far is updated to 8 with p_7. Return to previous step, p_8 is popped from PQ. The kth_so_far is not less than $dr_{min}(p_8, Q)$ and $|A| = k$, therefore, HS is stopped and report point p_7 as the result of that query.

4   Experimental Evaluation

In this section, we conduct extensive empirical evaluation based on real world POI dataset to verify the superior of our proposed solution. All the algorithms are implemented in Java and running on a PC with Intel 2.13GHz CPU and 4GB memory.

4.1   Experimental Setup

Road Network. The road network dataset used in our evaluation is Beijing road network, which contains 106,579 road segment nodes and 141,380 road segments.

Point Set. We use Beijing POI dataset in our experiment as point set. This dataset contains more than 500K POIs. We align each POI onto the road network to get its network location.

4.2   Evaluation Approach

We proposes three search algorithms, i.e, *exhaustive search* search, FS and hierarchical search HS. The performance, i.e., IO cost and execution time, of these search algorithms is affected by several parameters, such as number of reference points (denoted as n), number of query points (denoted as m) and number of requested results (denoted as k). Table 2 lists the default value and range of all these issues we used throughout the experiments. It is noteworthy that we adopt the variable-control method in the experiments that in each experiment set only one issue is adjustable and the rest issues are fixed to their default values.

Table 2. Parameter Settings

Parameter	Default value	Range
Number of grid cells	300	25, 50, 100, 200, 300, 400
Number of query points	9	5, 7, 9, 11, 13, 15
Number of requested results	10	5, 10, 15, 20, 25, 30

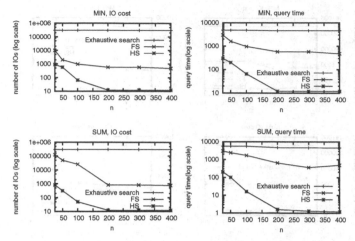

Fig. 3. Effect of number of grid cells

4.3   Evaluation Results

Effect of the Number of Grid Cells In this set of experiments, we evaluate how the performance of the algorithm is affected by number of grid cells. As shown in Figure 3, for each algorithm, the performance to solve min and sum aggregate function is similar. For both FS and HS algorithm, the IO cost and time cost decrease rapidly with number of grid cells growing from 25 to 200. The reason is that the more grid cells (i.e. more reference points) are deployed on the road network, the more closely query point q is to reference point r, which leads that $dr(p,q)$ is closer to $d_n(p,q)$. The closer $dr(p,q)$ to $d_n(p,q)$, in other words, is a tighter upper bound. Thus the performance of algorithms are improved very much. On the other hand, after deploying more than 200 grid cells, the increasing rate of performance becomes smoothly. This is because once the number of grid cells have reached a certain amount, the influence of adding more reference points is weak.

The last algorithm HS dominates FS under all value of n in both IO cost and query time aspects. This is because HS uses both hierarchical cluster structure and priority queue to prune unnecessary clusters during searching while FS not. Thus, the IO cost of HS is very low since it only need to search few clusters to find the answer of AkFN query, which is verified in this experiment. Meanwhile, this experiment also illustrates that using priority queue to sort the maximum aggregate distances can reduce query time obviously.

Effect of the Number of Query Points. Fig 4 demonstrates how the number of query points affects the effect of these three searching algorithms. We compare the three algorithms by tuning the number of query points from 5 to 15 with the step of 2. The IO cost of exhaustive search algorithm is same with previous experiment, which is still kept at a high level. The query time of exhaustive search algorithm increases rapidly due to calculating the aggregate distance becomes

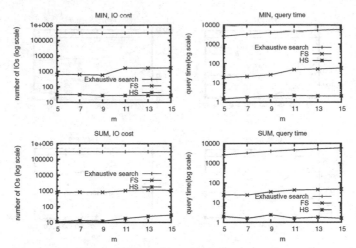

Fig. 4. Effect of number of query points

more complex when the number of query points is bigger. The FS algorithm is stable for *sum* aggregate function, but grows fast on *min* aggregate function. This is because the *min* aggregate function needs to find the maximum minimum aggregate distance while the *sum* aggregate function not, which means the bound of *min* aggregate function is looser than the *sum* aggregate function. HS shows the best performance for both the *min* and the *sum* aggregate functions in both IO cost and query time aspects. The reason is that the clusters in priority queue are sorted by maximum aggregate distance, thus most of unnecessary hierarchical clusters are pruned before HS searching in it. Hence no matter how many query points are given, it will not affect the efficiency of HS very much.

Effect of the Number of Requested Results. In this set of experiments, we evaluate how the performance of the algorithm is affected by number of requested results. The result is shown in Fig 5. We compare the IO cost and the query time of these three algorithms by tuning the number of requested results from 5 to 30 with the step of 5. The IO cost and query time of exhaustive search algorithm is similar with last two experiments, since it need to search the whole dataset. The performance of FS is stable except IO cost in *min* aggregate function, the reason is that the *min* aggregate function needs to find the maximum minimum aggregate distance. The IO cost of HS for both *min* and *sum* aggregate functions increases quickly, since HS needs to search more clusters in hierarchical structure to get top-k results, but its IO cost increases stable with the growing of the number of the requested points. However, the query time of HS is very stable with different k value since the HS maintains the hierarchical cluster structure and the priority queue. Hence HS can quickly get top k records from that priority queue during searching the hierarchical cluster. Finally, HS still outperforms other two algorithms in this experiment set.

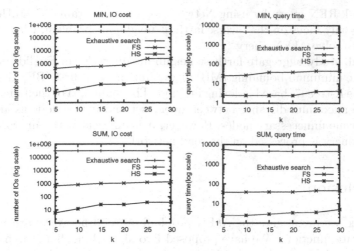

Fig. 5. Effect of number of requested results

5 Related Work

The problem investigated in our paper is combined with aggregate function and farthest neighbor search. We review the previous work related to these two categories in this section.

5.1 Aggregate Nearest Neighbor (ANN) Query

Papadias et al. [3] studied earlier version of ANN query, which is called GNN query. Three algorithms, which are called MQM, SPM and MBM, are proposed in their work to solve GNN query. Papadias et al. extended their work [5] and applied these algorithms to ANN query for solving *min* and *max* aggregate functions. There are several different types of GNN query have been recently studied in [10]. Yiu et al. studied the ANN query in road networks [2] from Papadias et al [1]'s work. They proposed three algorithms, which are called IER, TA and CE, to process *sum* and *min* aggregate functions with all data objects are index by R-tree.

All the above work focus on NN query, which are not applicable to our aggregate farthest neighbor query settings, where our query target to the farthest neighbor on the road network.

5.2 Farthest Neighbour (FN) Query

Yao et al. defined reverse farthest neighbour (RFN) query [11] problem. They proposed furthest Voronoi diagram based algorithms, which are called progressive farthest cell (PFC) algorithm and convex hull farthest cell (CHFC) algorithm to process RFN query with R-tree indexing. Moreover Tran et al. [12]

studied top-k RFN query by using Network Voronoi Diagram (NVD),However, these works only focus on the reverse farthest neighbour problem, where are not applicable to solve AkFN query.

Gao et al. study aggregate farthest neighbour query [6] in Euclidean space, and they the minimum bounding (MB) algorithm and best first (BF) algorithm, which all use R-tree [7] based indexing method. The main idea of their algorithms is using the maximum distance between query set and R-tree node as an upper bound to prune unnecessary nodes. However, it is different with our work, since our AkFN query is processed on the road network.

6   Conclusion

In this work, we investigate the AkFN problem on road network with min and sum aggregate functions. We have proposed two algorithms, flat search and hierarchical search, to associate with hierarchical clustering and pre-computing methods. Finally, the experimental results show that the flat search algorithm and the hierarchical search algorithm boost the searching efficiency compared with naive algorithm.

Acknowledgement. This research is partially supported by Natural Science Foundation of China (Grant No.61232006), and the Australian Research Council (Grants No. DP110103423, No. DP120102829 and No. DE140100215).

References

1. Papadias, D., Zhang, J., Mamoulis, N., Tao, Y.: Query processing in spatial network databases. In: PVLDB. VLD 2003, pp. 802–813 (2003)
2. Yiu, M., Mamoulis, N., Papadias, D.: Aggregate nearest neighbor queries in road networks. TKDE 17(6), 820–833 (2005)
3. Papadias, D., Shen, Q., Tao, Y., Mouratidis, K.: Group nearest neighbor queries. In: ICDE, pp. 301–312 (2004)
4. Cheong, O., Su Shin, C., Vigneron, A.: Computing farthest neighbors on a convex polytope. Theoretical Computer Science 296(1), 47–58 (2003)
5. Papadias, D., Tao, Y., Mouratidis, K., Hui, C.K.: Aggregate nearest neighbor queries in spatial databases. ACM Trans. Database Syst. 30(2), 529–576 (2005)
6. Gao, Y., Shou, L., Chen, K., Chen, G.: Aggregate farthest-neighbor queries over spatial data. In: Yu, J.X., Kim, M.H., Unland, R. (eds.) DASFAA 2011, Part II. LNCS, vol. 6588, pp. 149–163. Springer, Heidelberg (2011)
7. Guttman, A.: R-trees: A dynamic index structure for spatial searching. In: SIGMOD 1984, pp. 47–57 (1984)
8. Yiu, M.L., Mamoulis, N.: Clustering objects on a spatial network. In: SIGMOD 2004, pp. 443–454 (2004)
9. Jagadish, H.V., Ooi, B.C., Tan, K.L., Yu, C., Zhang, R.: idistance: An adaptive b+-tree based indexing method for nearest neighbor search. ACM Trans. Database Syst. 30(2), 364–397 (2005)

10. Xu, H., Li, Z., Lu, Y., Deng, K., Zhou, X.: Group visible nearest neighbor queries in spatial databases. In: Chen, L., Tang, C., Yang, J., Gao, Y. (eds.) WAIM 2010. LNCS, vol. 6184, pp. 333–344. Springer, Heidelberg (2010)
11. Yao, B., Li, F., Kumar, P.: Reverse furthest neighbors in spatial databases. In: ICDE, pp. 664–675 (2009)
12. Tran, Q.T., Taniar, D., Safar, M.: Reverse k nearest neighbor and reverse farthest neighbor search on spatial networks. In: Hameurlain, A., Küng, J., Wagner, R. (eds.) Trans. on Large-Scale Data- & Knowl.-Cent. Syst. I. LNCS, vol. 5740, pp. 353–372. Springer, Heidelberg (2009)

OSSM: The OLAP Security Specification Model

Ahmad Altamimi and Todd Eavis

Concordia University, Montreal, Canada
a_alta@encs.concordia.ca, eavis@cs.concordia.ca

Abstract. Security policies in Online Analytical Processing (OLAP) systems are designed to protect sensitive data from unauthorized access while, at the same time, ensuring that legitimate requests can be consistently satisfied. Ultimately, such policies allow administrators to define a series of restrictions and/or exceptions that can be associated with the components of the OLAP data model, including elements such as dimensions, cells, and aggregation hierarchies. A primary limitation of many current systems is that security policies are generally constructed on top of very granular privilege models that can produce complex and error prone mappings to the elements of the OLAP domain. In this paper, we present an Object Oriented Security Model (OSSM) that has been specifically designed for the specification of security policies within OLAP environments. In addition to explicit support for components of the conceptual data model, the OSSM can be used by the associated security policy engine to transparently and consistently propagate constraints across all relevant levels of dimension hierarchies. We discuss the core elements of the OSSM, as well as the integration with the policy engine that supports the language interfaces.

Keywords: OLAP, Security Languages, Authorization policy.

1 Introduction

One of the most important features of any OLAP system is the protection of data against unauthorized disclosure (privacy), while at the same time ensuring accessibility by authorized users whenever needed (availability). Considerable effort has been devoted to addressing various aspects of privacy and availability. Two main objectives are considered in this context. The first is the identification and specification of suitable security policies. The second is the development of a suitable access control mechanism implementing the stated policies.

With respect to the former, a number of researchers have investigated more powerful access control systems, including those designed specifically for OLAP domains [1,2]. Policy specification has also been considered though, in this case, the target has typically been large scale distributed computing and network environments [3,4]. We note that existing approaches to specification are not as efficient as those natively developed for OLAP. Specifically, policies in such environments define access privileges that are correlated with physical objects

H. Wang and M.A. Sharaf (Eds.): ADC 2014, LNCS 8506, pp. 26–37, 2014.

such as tables, files, or servers. In contrast, policies in OLAP domains are ultimately associated with abstract conceptual entities such as dimensions and hierarchical aggregation levels. Mapping these conceptual entities to the physical elements of the storage layer (i.e., tables, rows, columns) can, however, be a significant technical challenge for administrators. Furthermore, the existence of aggregation hierarchies provides opportunities for malicious users to subvert the intended protection layers. For example, one might work around a restriction on the summation of provincial sales totals by "rolling up" municipal sales results instead. Failure to protect all such possibilities would introduce potential vulnerabilities into the system.

In this paper, we propose a policy specification model that borrows from the feature set of the object-oriented paradigm. The OLAP Security Specification Model (OSSM) relies on the concepts of classes and objects to create instances of various policy constructs such as Subjects, Roles, and Protected Objects. These instances/objects are then combined together to create more expressive policies. A primary objective of this approach is to allow policy designers to identify security/privacy constructs at the level of the conceptual data model, without regard for the complexity of the underlying logical or physical implementation.

To support the evaluation of the OSSM approach, we provide a prototype implementation developed specifically for this environment. The framework consists of three main components. The User Interface Tool allows end users to design policies either by using an Object Oriented programmatic API or through a more conventional SQL-style interface that directly interacts with the DBMS. The Policy Repository, in turn, stores the policies generated by the interface tool. Finally, the Policy Manager retrieves policies from the repository and delivers them to the access control module for enforcement.

The remainder of the paper is organized as follows. In Section 2, we present an overview of related work. Section 3 discusses the terminology relevant to the policy specification domain. The OSSM model is then presented in detail in Section 4. The implementation framework, including language/API mappings is discussed in Section 5 and Section 6. Final conclusions are offered in Section 7.

2   Related Work

Early work in the area of policy specification tended to focus on networked and distributed environments. The Ponder model, for example, targets networked domains and represents policies as entries in a table consisting of multiple attributes [5]. This model was extended to fully distributed environments with Ponder2, an XML-based language that specifies security and management policies in a subject-action-target (SAT) format [6]. SecPAL, on the other hand, expresses security credentials using predicates defined by logical clauses, in the style of constraint logic programming [4]. We note, however, that while these approaches are efficient when used on large-scale networks and distributed systems, they are not well-suited to OLAP domains as they are often fragmented, dependent on infrastructure, and lack any native understanding of OLAP's multidimensional data model.

The ubiquitous Unified Modelling Language (UML) has also been used to formulate security policies. For example, Alam et al. propose a security language called SECTET-PL, and show how to express trust policies by using predicate expressions whose grammar is expressed in UML [7]. In [8], the authors present a Trust Management Framework that supports policy life cycle management using UML diagrams. However, none of this work specifically focuses on the OLAP domain; instead, it targets the policy model design itself.

Policy specification in web-based applications has also been proposed. SELinks, for instance, targets web apps and provides a uniform programming model (in the style of LINQ and Ruby on Rails), with language syntax for accessing objects residing either in the database or at the server [9]. Other frameworks investigate the association of security policies with client side code, with protection provided by the interception and analysis of queries [11].

Modifications to SQL have been discussed as well, including extensions to the SELECT and REVOKE statements in order to define a purpose driven authorization model [12]. SQL to XML transformations have also been studied. It is possible, for example, to employ control expression languages to map policies from relational environments to those represented in XML [13]. Apart from the fact that the model has not been implemented and its efficiency is unknown even for small databases, the model is not applicable to multidimensional models where millions of records may be materialized.

Finally, we note that while language extensions to specifically support OLAP have not been addressed, a number of researchers have investigated more general design issues for the data warehousing context, including both early requirement targets such as agents, decisional goals, and quality goals [2], as well as late stage conceptual-to-logical model mappings for authorization and auditing purposes [14]. A survey of objectives, features, and limitations of warehouse security modelling was provided in [15].

3 Preliminaries

Before discussing the semantics and syntax of the OSSM, we first provide an introduction to the conceptual data model upon which OSSM is based and give a brief overview of the basic terminology and structures relevant to policy specification in general.

3.1 The Conceptual Data Model

We consider multidimensional environments to consist of one or more *data cubes* [16]. Each cube is composed of a series of d dimensions — sometimes called *feature* attributes — and one or more *measures*. The dimensions can be visualized as delimiting a d-dimensional hyper-cube, with each axis identifying the members of the parent dimension (e.g., the months of the year). Cell values, in turn, represent the aggregated measure of the associated members. Figure 1 provides an illustration of a very simple three dimensional cube on *Store*, *Time* and *Product*.

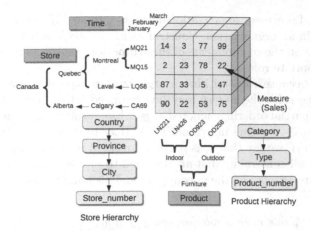

Fig. 1. A Simple three dimensional data cube

Note, as well, that each dimension can be associated with a distinct *aggregation hierarchy*. Products, for instance, are organized in Category → Type → Product_number groupings. Product_number is the lowest or *base* level in the Product dimension. In practice, data is physically stored at the base level so as to support run-time aggregation to coarser hierarchy levels. In this sense, we may consider the attributes of each dimension to be partially ordered by the dependency relation $\preceq$ into a *dependency lattice* [17]. We could therefore define the hierarchical relationships within the Product Dimension as Product_number $\preceq$ Type $\preceq$ Category. The dependency lattice is formally expressed in Definition 1.

Definition 1. *A dimension hierarchy H_i of a dimension D_i, can be defined as $H_i = (L_0, L_1, \ldots, L_j)$ where L_0 is the lowest level and L_j is the highest. There is a functional dependency between L_{h-1} and L_h such that $L_{h-1} \preceq L_h$ where $(0 \leq h \leq j)$.*

3.2 Subjects, Objects, Roles, and Policies

Security considerations in the data cube context range from simple authentication to the complex data authorization that provides protection for sensitive data. Such considerations can be achieved by using *policies*. Policies determine which subjects or users have access to a specific data object. For example, the policy could establish that a user is restricted from accessing a specific level of aggregation within a dimension but not the coarser levels.

Policies are typically based upon a combination of three basic components. The first, *Subjects*, represents users to which authorizations are granted. A Subject can be a single user or a group of users within the system. *Objects*, on the other hand, refer to the data to be protected. An object can be any partition of a data cube defined along one or more dimensions in order to give an additional opportunity for finer authorization. Finally, *Roles* are named collections of privileges and represent organizational agents intended to perform certain job functions within an organization. Constraints are subsequently assigned to specific roles based on the requirements of these functions. Subjects in turn are then assigned appropriate roles.

Ultimately, because roles within an organization typically have overlapping permissions, the roles themselves may be organized into a hierarchy which, in turn, defines a partial ordering, denoted as $\preceq$. More formally, given a role domain R, we let r_i, $r_j \in R$ define individual roles. If r_i precedes r_j in the hierarchy ordering $(r_i \preceq r_j)$, we say that r_i is partially ordered relative to r_j. This implies that r_i inherits all constraints that are assigned to r_j, and that all users who are mapped to r_i are affected by the r_j constraints. This notion is formally expressed in Definition 2.

Definition 2. *A role r_i in a role hierarchy R inherits all constraints of roles L = $(r_j, \ldots, r_z)$, where $r_i \preceq r_j$ and $r_j \preceq r_x \preceq r_z$ for a role $r_x \in R$. We say that r_i inherits all constraints of roles reachable from r_i to the Root role of R.*

4   The OLAP Security Specification Model (OSSM)

OSSM is a high-level OLAP security specification model designed to support the development of intuitive policy schemas mapping directly to the OLAP domain. The formal semantics for OSSM are based on the well-established typing and inheritance features of the Object Oriented paradigm. Specifically, OSSM defines basic policy components as a set of cooperative classes. Each such class focuses on a specific concept and defines all related functionality, including object instantiation and management, property specification, core operations and associated data structures. Objects can be re-used and combined in order to represent more sophisticated policies. The core OSSM classes are the Subject Class (SC), Object Class (OC), Role Class (RC), and Policy Class (PC). Before discussing the classes themselves, we first introduce several supporting definitions.

Definition 3. *Let A be a set of attributes shared by the objects of a specific class. A is defined as the union of the object's identifier OI and the object's attributes OA, where OA is a n-tuple of fields $(a_1, a_2, \ldots, a_n)$ that characterizes the object, and OI is an identifier that explicitly defines every object of that particular class for its entire life. We say that $A = OI \cup OA$.*

Definition 4. *For any protect object O, M is a set of methods allowed on O. These methods are classified into two broad categories: Control Methods that are used to create, update, and destroy objects, and Manipulation Methods that are used to maintain objects memberships and obtain objects information.*

Definition 5. *Let C be a set of conditions that specifies protected values. C is defined as an n-tuple $(c_1, \ldots, c_n)$ that may be connected by logical operators (AND, OR) to define complex predicates protecting a cube element. We say that C forms a general representation of any criteria/condition that may restrict any element in a data cube by applying both arithmetic and logical operators.*

Definition 6. *Let E be a set of conditions (e_1, ..., e_m) that specifies exception values. E defines a subset of the protected values as an exception, such that $E \subset C$.*

We will use following example throughout this section to illustrate the functionality of the above classes.

Example 1. Given the data cube depicted in Figure 1, we assume the the following two restrictions should be satisfied. Due to privacy concerns, the sales totals of all cities in the province of Quebec should not be accessed by the user Bob, except for the sales of the city of Montreal. Second, assume that any sale completed before 2005 should not be used for analysis.

4.1   The Subject Class (SC)

The Subject Class (SC) provides a template, or blueprint, to define the properties and the operations common to all of the system's users. These properties and operations are represented within a class as fields and methods, and are defined as a tuple (A, M). By applying the previous definitions to Example 1, we have a Subject (e.g., *Bob*) with the following elements:

- A = $OI \cup OA$, where OI is the Bob identifier, and OA is the set of attributes that describes Bob (e.g., subject name, password, and role identifiers).
- M = A method set (e.g., createSubject(), updateSubject(), dropSubject()).

4.2   The Object Class (OC)

The Object class (OC) provides a template to describe the protected data. We note, however, that while the information within the cube is aggregated at distinct levels of granularity, the underlying physical data may in fact be distributed across multiple tables. Moreover, we must be aware that protected data can be computed from its more granular levels. Thus, any level below the protected data must also be identified as protected. As such, the OC class must expose the conceptual features of the OLAP data model, while transparently mapping these entities to the logical schema. In practice, OSSM utilizes the XML-based grammar of the open source Mondrian OLAP server for the conceptual-to-relational mapping [18]. Note that this mapping is a one-time operation (carried out by the database schema designer) and is entirely hidden from the policy administrator.

We now turn to the declaration of the OC itself. The OC consists of a tuple (A, C, E, M), where A is a set of attributes describing the data object to be protected, C is a set of conditions that specifies the protected values, E is a set of conditions that specifies the exception, if one exists, and M is the set of methods allowed on the class objects. For instance, recall the restrictions of Example 1, whereby the sales of Quebec cities are restricted except for the city of Montreal. A protected object *Protected-Stores* might be defined as follows:

- $A = OI \cup OA$.
- C = The restriction (e.g., Stores.Province = "Quebec").
- E = The exception (e.g., Stores.City = "Montreal").
- M = A set of methods (e.g., createObject(), dropObject(), updateObject()).

4.3   The Role Class (RC)

A *role* regulates the activities of its members through a set of restrictions. Instead of specifying such restrictions for each user, the OSSM specifies constraints via data objects representing shared roles. In other words, restrictions are associated with the data objects that are grouped together or encapsulated into Role objects, as formally stated in Definition 7.

Definition 7. *Let OC_1, OC_2, ..., OC_n be n objects of Object Class (OC). We define a Role class from these n objects as a tuple (O,A,M), where O is the set of protected objects created by OC.*

Example 2. Suppose a role (e.g., *Analysis*) is defined with the protected object *Protected-Stores* . The Analysis role would be defined as:

- O = {*Protected-Stores*}.
- A = $OI \cup OA$, where OI is the role identifier, and OA is the set of attributes to describes the role (e.g., role name, role description).
- M = A set of methods (e.g., createRole(), dropRole(), addObjects()).

4.4   The Policy Class (PC)

We now proceed to the definition of the composite policy class (PC). The PC allows administrators to specify the requirements of the organization's security policies relative to the various cube elements. These requirements outline the association between Subjects, Protected Data Objects, and Roles. We formally define of PC below, and then provide an example to illustrate its construction.

Definition 8. *Let SC be a Subject Class and RC be a Role Class. We define a Policy Class (PC) as a tuple (SO,RO,A,M), where SO and RO are the Subject and Role objects that have been instantiated from SC and RC.*

Example 3. By applying the class definition of PC to the policy of Example 1, the following policy (e.g., *Policy1*) can be constructed.

- SO = {*Bob*} — the subject.
- RO = {*Analysis*} — the protected object.
- A = Policy Identifier $\cup$ an attribute set, used to describe the policy and the association between its components (e.g., role name, role description, Subjects Roles Assignment).
- M = A set of methods (e.g., Assign(), Withdraw(), and Select()).

5   The OSSM Policy Engine

While the proposed security model provides the logic and syntax for defining security policies, it is important to note that without an appropriate server-side engine to support the client-side language interfaces, the process of defining, storing, and managing security policies would not be possible. In this section, we discuss the OSSM policy management engine and provide a brief description of its structure and functions. The engine itself consists of three major components:

1. **The User Interface** is the means by which an administrator actually defines policies. It is an important component of the framework since it hides low-level policy details and permits the administrator to define the policy classes using either a declarative syntax or an OOP interface. Details of the options for language interfaces are discussed in Section 6.
2. **The Policy Repository** is used to store the policies generated by the language interface. It acts as a bridge between the interface software and the policy management component. The Policy Repository itself consists of a set of tables (Subjects, Objects, Roles, and Policies) that collectively represent the meta data required to define security measures. A slightly simplified depiction of its schema is provided in Figure 2(a).
3. **The Policy Manager** is a bidirectional component. The access control module requests policies that are associated with a specific user when it receives his/her query. The policy manager, in turn, contacts the policy repository and retrieves the applicable policies, which are then sent to the access control module that ultimately permits or rejects requests for a particular data item. The basic architecture is depicted in Figure 2(b).

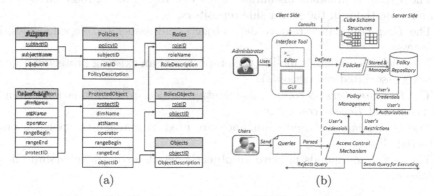

(a)   (b)

Fig. 2. (a) The Policy Repository (b) The Policy Engine

6   Integration Options

Policies that have been specified as per the OSSM model must eventually be entered into a policy repository in some way. Typically, this would be done in two distinct, but cooperative forms. On the one hand, the conceptual representation can be exposed directly within an Object Oriented programmatic API. On the other, a more conventional SQL-style interface can be provided for direct interaction with the DBMS in a manner consistent with conventional approaches. We note that the current prototype contains proof-of-concept versions of both. Because the OO approach ultimately "piggy backs" on top of the SQL model, we begin the discussion of the integration options with this more conventional implementation.

6.1   Declarative Language Extensions

Existing access policies are typically specified via SQL statements, most notably the GRANT/REVOKE commands of the Data Control (sub) Language (DCL). Here, basic privileges (e.g., INSERT, SELECT, UPDATE, EXECUTE) can be associated with the elements of the logical schema (e.g., tables, views, procedures). For convenience, privileges can in turn be organized into simple Roles. For example, the **testing** Role can be created by the database designer, with the ability to create new tables then assigned to this role.

The more sophisticated security constraints described in this paper can in fact be integrated into this specification environment. In particular, the OO characteristics of the conceptual model can be "flattened" to suit the simple declarative structure of the DCL without sacrificing expressibility. Below, we list the core *command categories* that can be used to extend the conventional SQL/DCL language. Figure 3 then illustrates basic syntax, along with a series of simple examples.

- The *Create* commands instantiate Subjects, Roles, and Protected Objects and store them in the server side repository.
- The *Drop* commands remove data objects and associated dependencies, if necessary.
- The *Update* commands modify the properties of an existing protected object or its associated restriction(s).
- The *Role membership* commands maintain subject-role information through *Assign* and *Revoke* commands.
- The *Object manipulation* commands are used to manage the policy's protected objects.
- The *Select* commands retrieve an object's details or its membership information.

6.2   Programmatic API

While SQL has been the de facto standard for general database interaction for the past several decades, programmatic APIs have also been developed. Perhaps most significantly, ODBC and JDBC have become the standard means by which to deliver query statements to the DBMS and, subsequently, receive results. More

The Command	The Command *Syntax* and Examples	Description
Create Subject	*Syntax: Create Subject userName With password* Example: Create User Sue with SOK92	Creates a new user account identified by a password.
Create Role	*Syntax: Create Role roleName Child? parentName?* Example: Create Role Marketing Child Administering	Defines a new role that can be a child of another existing role.
Create Restriction	*Syntax: Create Restriction restrictName On cubeElement* *Except exception* Example: Create Restriction RegionRest On Customer.Region Except Customer.Nation	Creates a restriction on accessing a specified data cube element with exception(s).
Drop Subject	*Syntax: Drop Subject subjectName* Example: Create User Sue with SOK92	Removes a user account permanently. Consequently, its roles memberships will also be removed automatically.
Drop Role	*Syntax: Drop Role roleName* Example: Drop Role Marketing	Removes a specific role. Consequently, any user membership for the specified role is revoked.
Drop Restriction	*Syntax: Drop Restriction restrictionName* Example: Drop Restriction RegionRest	Removes a defined restriction and consequently removes it from all associated roles.
Update Restriction	*Syntax: Update restrictionName Set Restriction newRestriction* Example: Update RegionRest Set Restricion Customer.Region=Asia	Updates the restriction itself.
Update Exception	*Syntax: Update restrictionName Set Exception newException* Example: Update RegionRest Set Exception Customer.Nation=China	Updates the Exception if it is exits.
Assign	*Syntax: Assign subjectName To roleName* Example: Assign Sue To Marketing	Assigns an existing subject to one or more existing roles according to the subject duties.
Revoke	*Syntax: Revoke subjectName From roleName* Example: Revoke Sue From Marketing	Revoke a subject from a specific role.
Add	*Syntax: Add restrictionName To roleName* Example: Add RegionRest To Marketing	Adds a restriction to a specific role. All subjects assigned to that role will be affected by this restriction.
Remove	*Syntax: Remove Restriction restrictionName From roleName* Example: Remove Restriction RegionRest From Marketing	Removes a restriction and/or its exception from a specific role to make it less restrictive.
Select Subjects	*Syntax: Select Subjects Of Role roleName* Example: Select Subjects Of Role Marketing	Gets all subjects assigned to a specific role.
Select Roles	*Syntax: Select Roles Of Subject subjectName* Example: Select Roles Of Subject Sue	Gets all roles of a specific subject.
Select Subject's Restrictions	*Syntax: Select Restrictions On Subject subjectName* Example: Select Restrictions On Subject Sue	Gets all restrictions on a specific subject.
Select Role's Restrictions	*Syntax: Select Restrictions Of Role roleName* Example: Select Restrictions Of Role Marketing	Gets all restrictions of a specific role.

Fig. 3. Delarative OSSM commands

recently, Object Relational Mapping (ORM) frameworks such as Hibernate [19] have been developed to minimize the impact of the impedance mismatch caused by Object-to-Table mapping logic.

In such cases, of course, it is important to note that the queries encapsulated by the API methods are typically data queries. In other words, such queries are interactively retrieving and updating dynamic operational data. Strictly speaking, it is possible for an OO implementation of the security classes previously discussed to be used directly against the DBMS. However, in most situations it would be cumbersome to specify and maintain policy specifications programmatically (i.e., writing application code to view and maintain policy objects). In fact, a more likely scenario would be the use of graphical tools to allow intuitive modelling and maintenance of complex enterprise policies.

It is in this context that an OO API would be utilized. Specifically, the design of modelling applications would be significantly simplified by a direct proxy interface. Here, the policy classes discussed in the preceding sections are exposed as wrappers to the backend repository. Policy objects would then be accessible within the graphical interface so as to provide security specialists with an intuitive, point-and-click mechanism for setting and modifying authorization constraints.

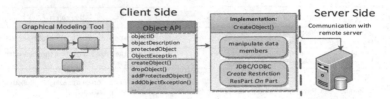

Fig. 4. OSSM API logic

Figure 4 provides a simple illustration of the model. Here a client side Object (i.e., OC) houses data related to a given Policy Object specification. The implementation exposes an OO API to the privacy specialist who can then manipulate the object graphically. It terms of the class implementation itself, it would function as a proxy in the sense that it would relay data (e.g., object instantiation and updates) to the DBMS repository. To do so, it would utilize the low level JDBC/ODBC API to transmit the extended SQL statements described in the previous subsection. All of this logic would be transparent to the user as he/she would simply work with the graphical representation of the policy objects.

7   Conclusions

In this paper, we have introduced a security policy design model called OSSM that is directly associated with the OLAP domain. Building on an Object Oriented paradigm, OSSM allows administrators to intuitively model complex OLAP-specific policies at the level of the conceptual data model. As a consequence, the multi-level aggregation hierarchies inherent in the data model can be secured without concern for the complexities or idiosyncrasies of the underlying logical or physical schema. Subjects, Roles and Protected Objects can be combined and re-used as required. We have also discussed the associated policy engine, as well as options for integrating the OSSM approach into conventional data management systems. Given the size of the OLAP market, and the importance of properly protecting analytics/warehouse data, we believe that this kind of domain-specific approach represents a significant improvement relative to current alternatives.

References

1. Altamimi, A., Eavis, T.: Securing Access to Data in Business Intelligence Domains. J. International Journal on Advances in Security 5, 94–111 (2012)
2. Khajaria, K., Kumar, M.: Modeling of security requirements for decision information systems. J. SIGSOFT Softw. Eng. Notes 36, 1–4 (2011)
3. Dell'Amico, M., Serme, G., Idrees, M.S., Santana de Olivera, A., Roudier, Y.: HiPoLDS: A security policy language for distributed systems. In: Askoxylakis, I., Pöhls, H.C., Posegga, J. (eds.) WISTP 2012. LNCS, vol. 7322, pp. 97–112. Springer, Heidelberg (2012)

4. Becker, M., Fournet, C., Gordon, A.: SecPAL: Design and semantics of a decentralized authorization language. J. Comput. Secur. 18, 619–665 (2010)
5. Damianou, N., Dulay, N., Lupu, E.C., Sloman, M.: The Ponder Policy Specification Language. In: Sloman, M., Lobo, J., Lupu, E.C. (eds.) POLICY 2001. LNCS, vol. 1995, pp. 18–38. Springer, Heidelberg (2001)
6. Twidle, K., Dulay, N., Lupu, E., Sloman, M.: Ponder2: A Policy System for Autonomous Pervasive Environments. In: IEEE Workshop on Policies for Distributed Systems and Networks, pp. 330–335. IEEE Computer Society, Washington (2009)
7. Alam, M., Breu, R., Hafner, M.: Model-Driven Security Engineering for Trust Management in SECTET. J. Journal of Software 2, 47–59 (2007)
8. Halvard, S., Hamid, M., Boualem, B., Fabio, C.: Modeling Trust Negotiation for Web Services. J. Journal of Computer 42, 54–61 (2009)
9. Corcoran, B., Swamy, N., Hicks, M.: Cross-tier, label-based security enforcement for web applications. In: ACM SIGMOD, pp. 269–282. ACM, New York (2009)
10. Jacobi, I., Kagal, L., Khandelwal, A.: Rule-Based Trust Assessment on the Semantic Web. In: Bassiliades, N., Governatori, G., Paschke, A. (eds.) RuleML 2011 - Europe. LNCS, vol. 6826, pp. 227–241. Springer, Heidelberg (2011)
11. Felt, A., Finifter, M., Weinberger, J., Wagner, D.: Diesel: applying privilege separation to database access. In: 6th ACM Symposium on Information, Computer and Communications Security, pp. 416–422. ACM, New York (2011)
12. Van Staden, W., Olivier, M.: SQL's revoke with a view on privacy. In: South African Institute of Computer Scientists and Information Technologists on IT Research in Developing Countries, pp. 181–188. ACM, New York (2007)
13. Leighton, G.: Preserving SQL access control policies over published XML data. In: EDBT/ICDT Workshops, pp. 185–192. ACM, New York (2009)
14. Soler, E., Trujillo, J., Fernandez-Medina, E., Piattini, M.: Application of QVT for the Development of Secure Data Warehouses: A case study. In: Int. Conference on Availability, Reliability and Security, pp. 829–836 (2007)
15. Singh, I., Kumar, M.: Evaluation of approaches for designing secure data warehouse. In: Int. Conference on Advances in Computing, Communications and Informatics, pp. 69–73. ACM, New York (2012)
16. Jim, G., Adam, B., Andrew, L., Don, R., Hamid, P.: Data cube: A relational aggregation operator generalizing group-by, cross-tab, and sub-totals. J. Data Mining and Knowledge Discovery. 1, 29–53 (1997)
17. Harinarayan, V., Rajaraman, A., Ullman, J.: Implementing data cubes efficiently. In: ACM SIGMOD, pp. 205–216. ACM, New York (1996)
18. Mondrian. Pentaho Analysis Services, http://mondrian.pentaho.com
19. Linwood, J., Minter, D.: Beginning Hibernate, 2nd edn. Apress (2010)

Scalable Gaussian Process Regression for Prediction of Material Properties

Eve Bélisle[1], Zi Huang[1], and Aimen Gheribi[2]

[1] University of Queensland, Brisbane, Australia
[2] École Polytechnique de Montréal, Montréal, Canada
{uqzhuang,e.belisle}@uq.edu.au, aimen.gheribi@polymtl.ca

Abstract. Gaussian process regression (GPR) is a non-parametric approach that can be used to make predictions based on a set of known points. It has been widely employed in recent years on a variety of problems. However the Gaussian process regression algorithm performs matrices inversions and the computational time can be extensive when accessing large training datasets. This is of critical importance when on-line learning and regression analyses are carried out on real-time applications. In this paper we propose a novel strategy, utilizing batch query processing and co-clustering, to achieve a scalable and efficient Gaussian process regression. The proposed strategy is applied to a real application involving the prediction of materials properties. Comprehensive tests have been conducted on two published properties data sets and the results demonstrate the high accuracy and efficiency of our new approach.

1 Introduction

The Gaussian Process Regression (GPR) is a well-known and highly reliable regression model in Machine Learning. Its non-parametric nature makes it flexible and particularly adaptable to various types of data. It has been widely used in scientific data analysis, such as prediction of materials properties, microstructure evolution prediction in thermo-mechanically processed metals, robot control, etc. For example, given a set of known chemical compounds with their corresponding measured Electrical Conductivity (EC), the EC of an unknown material of known chemical composition (ex. $x_1 SiO_2 + x_2 CaO + x_3 MnO + x_4 Al_2O_3$) can be predicted by GPR, where SiO_2, CaO, MnO, and Al_2O_3 are considered as the multiple variables in this regression model.

Our work illustrated in this paper is inspired by an application on the prediction of material properties, more specifically, by the optimisation of these predictions, which require a large number of single predictions to be performed sequentially. Each prediction request from users is considered as a query in our work, which is represented as a single vector of real numbers, corresponding to the values of composition for each input component. The result of a query is a predicted value for the studied property. Though GPR has proven to be superior to other existing regression models in terms of reliability, it suffers from high computational cost caused by matrix inversion operations in both the learning

H. Wang and M.A. Sharaf (Eds.): ADC 2014, LNCS 8506, pp. 38–49, 2014.

and regression steps. In some cases, the learning step is only required to be pre-
formed once, as the learned hyperparameters of the model can be repeatedly
used for subsequent queries. However, applications such as material property
predictions are generally for more than one query. Scientists may upload a large
number of chemical compounds with different x_i constraints in order to make
EC predictions. The low efficient regression step in the conventional GPR is not
capable of dealing with the streaming queries on large scale. For a growing number of real-time applications such as robot dynamic control, on-line learning is
also required. It is extremely time consuming when applying the conventional
GPR, which makes real-time responses impractical.

In this paper, we propose a novel approach to perform the conventional GPR
efficiently with a three-step strategy. With this so-called Scalable GPR, the size
of the training data used for learning and regression is significantly reduced,
resulting in a promising efficiency improvement. Meanwhile, the intrinsic information embedded in the training data is kept in the reduced data set, which
guarantees a high accuracy of the regression. Two real applications are studied
in this paper: prediction of Martensite start temperature and Electrical Conductivity. The comprehensive experiments on these two materials datasets show
the outstanding performance of the proposed method compared with the conventional GPR and other existing popular machine learning methods for predictions.
To be more specific, we make the following contributions.

- We propose a fast batch query processing algorithm to handle large numbers
  of queries by grouping them by similar characteristics. It is an essential step
  for the real-time predictions and also the foundation of the further training
  data condensation.
- We analyze the structure of the training data and condense it by removing
  the redundant information and preserving embedded intrinsic information.
- A query-aware training data selection strategy is designed to further enhance
  the efficiency of the model by taking into account the relationship between
  the query and the training data.
- We conduct extensive performance studies on two real-life materials datasets,
  which are large scale from the perspective of machine learning. The results
  demonstrate the high accuracy and the significant efficiency improvement of
  our proposal over existing methods.

The rest of the paper is organized as follows. After discussing the related work,
we present a brief definition of the Gaussian Process Regression method. We
then describe the strategy developed in order to reduce the computational cost,
followed by the performance study on both Martensite Temperature prediction
and Electrical Conductivity prediction. Finally, a general discussion and analysis
of the results is presented.

2   Related Work

Predicting the martensite start temperature (Ms) has been reported by several
authors. While some had good results using a neural network model [14] [10],

others preferred a thermodynamic framework [11] or a purely empirical approach [4][5]. These methods have been thoroughly investigated by Soumail et al. in 2006 [9]. Their conclusion was that although the thermodynamic approach provides satisfying results, there is a strict limitation in the query points, based on the fundamental assumptions upon which the model was based. They found that the neural network approach performs as well as other methods, however some wild predictions were obtained and they recommended the use of a Bayesian framework.

The problem of high-dimensionality and large amount of data for Gaussian processes has been studied by E. Snelson et al. [8] and R. Urtasun [12]. They both proposed partitioning the data, which is the approach we adopt in this present study.

In the area of clustering of high-dimensional and large amount of data, Huang et al. [3] introduced an effective co-clustering approach, this method was used for multimedia similarity search and was not fully compatible with databases containing chemical compositions but we took inspiration from both ideas.

3   Gaussian Process Regression

In this section we give a brief description of the Gaussian process regression approach for machine learning. A Gaussian process (GP) is a generalisation of the Gaussian probability distribution [6]. It is essentially an extension of multivariate Gaussian, which considers Gaussian distribution to be not only over random vectors but also over random functions. A stationary Gaussian process regression (GPR) is the canonical statistical model for data arising from computer experiments. Like other Bayesian regression algorithms, the GPR computes a posterior distribution based on a prior distribution (training data). The GPR has been recognised as a superior method for the accurate function approximation in high-dimensional space [6]. For a detailed description of the GPR, please refer to Rasmussen and Williams [6].

In this work we consider the the covariance matrix as in previous work of Gibbs and MacKay[2]:

$$\underline{\underline{K}}(\overrightarrow{X}, \overrightarrow{X'}) = \sigma_f^2 \exp\left\{ -\frac{1}{2} \sum_{j=1}^{n} \frac{(x_i - x_j')^2}{d_j} \right\} + \sigma_n^2 \delta(\overrightarrow{X}, \overrightarrow{X'}) \tag{1}$$

Where δ is the Kronecker delta function, σ_f^2 denotes the function variance and d represents the width of the Gaussian kernel.

From the above equation, we can see that to perform a GPR, we need to generate the covariance matrix $\underline{\underline{K}}$, whose elements are covariance functions on all possible combinations of training data point pairs. To calculate $\underline{\underline{K}}_*^T$ for the regression part, we also need to process all training data points. Thus, the computational cost of a GPR heavily depends on the training data size. Though the GPR is a highly accurate regression method, it suffers from a high computational cost and is not practical for real-time applications.

4 Our Approach

The standard GPR algorithm has two main components: the optimisation of the hyperparameters to be used in the covariance matrix and the actual regression with the query points. Both require matrices inversions, and the computational cost is therefore heavily linked to the size of the training database. Typically, the computational complexity of performing the necessary matrices inversions is proportional to n^3 where n is the number of training data points. A scalable GPR is highly desirable because of the following two issues. Firstly, when given a fixed training data set, the optimisation step only needs to be performed once, since the hyperparameters can be saved and reused. However, to achieve accurate predictions for different kinds of query points, the training set has to contain as much information as possible, which results in a very large scale training data set. For this reason, the existing methods generally suffer from loading the training set with large amounts of data points. Secondly, in many real-time applications, such as robot control, on-line learning and regression are required. Since the computational cost of GPR is highly associated with the training data size, in the present study, we aim to design a scalable GPR algorithm by reducing the training data size while maintaining the intrinsic information embedded in it. The proposed algorithm has three stages: 1 - Batch query processing, 2 - Training data condensation and 3 - Query-aware training data selection. In the following sections we refer to our strategy as the Scalable GPR.

4.1 Batch Query Processing

While typical materials optimisation calculations are performed sequentially as they are dependent on the previous result, we will be considering large amounts of input queries in our application. In order to reduce the computational cost on on-line regression for streaming queries, we conduct batch query processing by considering the similarities between query points. According to their different characteristics, the query points are clustered into groups, each of which will be represented by a summarized representative point. The representatives will be passed to the regression model and be used for the training data selection. We apply an agglomerative clustering approach to first group points in pairs of closest points using the Euclidean distance. It then groups pairs together and so on until a target number of points per group is obtained. The function used to measure the Euclidean distance between two points and two groups of points is as follows:

$$\sqrt{\sum_{i=1}^{n}(q_i - p_i)^2} \qquad (2)$$

Where p and q are two points in an Euclidean space of dimension n. This method for clustering data in high-dimensional space has proven to be a simple but efficient one [13].

When comparing two groups of points, the geometrical mean on each dimension is used to calculate the Euclidean distance. Given a data set $\{\vec{x_1}, \vec{x_2}, ...\vec{x_n}\}$, the definition of geometrical mean on dimension d is as follows:

$$\left(\prod_{i=1}^{n} x_{di} \right)^{\frac{1}{n}} \tag{3}$$

The geometrical mean of a set is based on the product of the values instead of a sum. This type of mean is particularly useful when attempting to minimise the impact of data with different ranges, which could occur with data for prediction of material, where different scales might be found in the set of points.

4.2 Training Data Condensation

The second step consists of a pre-filtering of the entire training data. This is done to condense redundant observations, and therefore acts as a first dimensionality reduction step. Inspired by the co-reduction approach introduced by Huang et al. [3], we reorganise the rows by similarity and then combine them together using a reduction function Θ. Our reduction function Θ consists of computing the mean values on each dimension of the two merged rows. A reduction on the number of columns will be achieved in the final selection of data (section 4.3). Here a simple Euclidean distance function between the points is not enough, because we want to avoid a situation where two points would be far on one dimension and identical in every other dimensions. We want the clustered points to be close to each other on every dimension, due to the nature of the data. With chemical compositions, it can be the case that one of the components makes a very big difference on the value of the physical property, as there could be possible interactions with the other components present. For this same reason, we could not fully apply the co-reduction technique and introduce a column reduction function. For example, let us consider the following three points in a 4 dimensional space:

$$
\begin{array}{c}
& \begin{array}{cccc} C & Si & N & Mo \end{array} \\
\begin{array}{c} A \\ B \\ C \end{array} &
\left(
\begin{array}{cccc}
20 & 3 & 0 & 1 \\
15 & 3 & 4 & 1 \\
26 & 5 & 0 & 4
\end{array}
\right)
\end{array}
$$

Using equation 5, the Euclidean distance between A and B gives a value of approximately 6.4, while the Euclidean distance between A and C gives a value of 7. According to our previous reasoning, we wish to favour the clustering of A and C because they have actual data in the same dimensions, thus reducing the risk of component interaction affecting the physical property. Therefore, we use the following rule to compare two points p and q:

$$\forall i \in \mathrm{N} : \left(\frac{|q_i - p_i|}{\sum_{j=1}^{N} q_j} < \epsilon \right) \wedge \left(\frac{|q_i - p_i|}{\sum_{j=1}^{N} p_j} < \epsilon \right) \wedge ((p_i = 0) \leftrightarrow (q_i = 0))$$

Where N is the total number of columns (dimensions) and ϵ is an arbitrary condensation constraint, we tested with values of 0.5, 1, and 5%. If the above predicate is true, then the two rows can be merged together applying Θ. The algorithm is executed recursively until no more merges are possible.

4.3   Query-Aware Training Data Selection

Before calculating the actual predictions, we perform the final selection of the training points by considering the relation between the representative query points generated from the first stage and the condensed training set created in the second stage. This is done by first calculating the geometrical mean (i.e., Equation 3) on each dimension of the batch query. Once the geometrical mean (g) is found, we compare this value to each point (p) in the condensed training set obtained, using a modified Euclidean distance formula:

$$\sqrt{\sum_{i=1}^{n}\left(\frac{g_i}{\sum_{j=1}^{n}g_j} - \frac{p_i}{\sum_{j=1}^{n}p_j}\right)^2} \tag{4}$$

In other words, we calculate the Euclidean distance on normalised values. It is because we want to measure the distance using proportions of chemical compositions instead of the actual values. We keep an arbitrary number of results in the final training set, the ones with the closest distance to the geometrical mean (Fig. 4.3). A number of K similar points from the training data set will be selected for each batch query to be considered as its local or specific training data. The parameter K for each batch query is determined by the acceptable predicted error bound. That means the value of K is decided depending on the accuracy of the prediction. As you can see on Fig. 4.3, some points can be present in more than one training set, this will ensure consistency for each batch query.

If the target error bound can not be reached using the selected number of training data, the number of training points is increased and the regression is calculated again until the target error bound is reached. The reduced number of training points allows us to do a further clustering of the data, eliminating the dimensions where there is no composition available, therefore reducing the number of columns in the training matrix.

5   Application on Martensite Start Temperature Prediction

Martensite is a crystalline structure formed in the process of cooling carbon steels at high rates (quenching). Controlling the amount of martensite in a given steel is critical as it has an important effect on the physical and mechanical properties of the steel. One of the variables engineers have to take into account is the Martensite Start (Ms) Temperature, which can be predicted by giving the amount of each chemical component contained in a query steel. To evaluate the performance of the proposed Scalable GPR, we conduct a series of experiments on Ms temperature predictions.

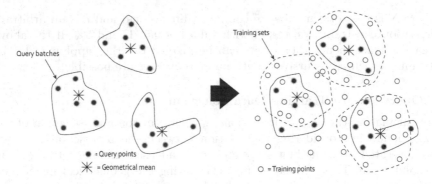

Fig. 1. Final selection of the training points: K-NN of the geometrical mean

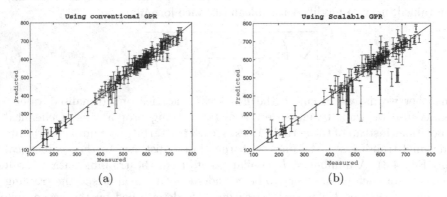

(a)   (b)

Fig. 2. Predicted vs Measured Ms Temperature (K)

5.1   Dataset

A database of approximately 1,100 entries collected from the literature, available for download on the Thomas-Sourmail website [10] is used in our experiment. It covers a wide variety of compositions of steels with measured Ms temperatures. Each entry represents a composition of the steel and its corresponding Ms temperature observed. The database is a table of 15 columns, where the first 14 columns represent the values in weight percent of 14 chemical components (C, Mn, etc.) and the last one is the Ms temperature value.

We randomly take 80% of the available points for training and the remaining 20% for testing the predictions. Thus, the numbers of training points and testing points are 870 and 220 respectively. This procedure is repeated 10 times to get the final prediction performance.

5.2   Performance Study

Conventional GPR: To illustrate the superiority of the proposed Scalable GPR, we first test the conventional GPR by using the training and testing data described in Section 5.1. The prediction values obtained in our experiment are

Table 1. Conventional GPR vs Scalable GPR for predicting Ms

	AE (%)	RMS (K)	Total training time (sec)	Average prediction time per testing point (sec)	Average time cost per testing point (sec)
GPR	3.08	21.6	969.40	5.52	9.92
Scalable GPR	5.02	42.6	26.5	0.13	0.25

Table 2. Batch query performances for prediction of Ms

Batch	Size of the training matrix	Training time for each batch query (sec)	Average prediction time per testing point (sec)
1	95×13	4.35	0.14
2	93×12	5.29	0.17
3	96×13	3.06	0.10
4	95×11	2.48	0.08
5	108×14	4.99	0.14
6	165×12	7.09	0.14

as consistent as those of Bailer-Jones, Bhadeshia and MacKay [1] (Fig. 2(a)). Two performance indicators average error (AE) and root mean square (RMS) are used to evaluate the accuracy of the testing method, which are defined as follows:

$$AE = \frac{1}{N_t} \times \sum_{i=1}^{N_t} \frac{|p_i - a_i|}{a_i} \tag{5}$$

$$RMS = \sqrt{\frac{\sum_{i=1}^{N_t} |p_i - a_i|^2}{N_t}} \tag{6}$$

Where N_t is the number of testing points, p is the predicted value and a is the actual value.

As reported in Table 1, the AE and RMS produced by the conventional GPR are 3.08% and 21.6 degrees respectively. The average prediction time in the regression step for each testing point is 5.52 seconds, on an Intel i7 3.4GHz with 16 GB of RAM and the time cost to calculate the hyperparameters in the training step is 969.40 seconds. As we mentioned earlier, on-line training is required in many real-time applications. Taking into account both the training and prediction time costs, GPR averagely spends 9.92 seconds on each testing point to make a prediction. The time cost at this scale is certainly impractical.

Scalable GPR: Our proposed Scalable GPR offers a significant efficiency improvement. Compared with GPR, the training time is heavily reduced from 969.4

seconds to 26.5 seconds and the average prediction time in the regression step is reduced from 5.52 seconds to 0.13 seconds. It makes the real-time prediction realistic where the total cost including both training and prediction for an individual query point is 0.25 seconds.

As described in section 4, batch query processing is performed in the Scalable GPR to achieve efficient query predictions. Here we choose one round of testing as an example to describe the batch query details. As illustrated in Table 2, a total of 6 batch queries are created in this testing round. We choose to set a condensation constraint of $\epsilon = 5\%$ which allows us to condense the training data from 870 points to 416 points in the second stage. The final selection of training points for each batch query is then performed using this condensed dataset, as explained in section 4.3. When trying to further condense the data, we observe that using an $\epsilon > 5\%$ leads to too much compression of the data, producing values in every dimensions for too many data points. In addition to a loss of information, this means that further vertical condensation is virtually impossible and therefore there is no further gain on the computational cost. In this example we set a target error of 10% or less, increasing the number of training points and using a lower condensation (ϵ) if not reached. The time cost to optimise the hyperparameters in the training step for each batch is reported in Table 2. We can observe that reducing the size of the training matrix is of critical importance to improve the speed of the GPR.

It is always a trade off between efficiency and accuracy. To achieve scalable and efficient predictions, the accuracy of the Scalable GPR is sacrificed, where the AE and RMS are 5.02% and 42.6 degrees respectively (Table 1). However, from a chemistry point of view, an average error of 10-20% or less is considered acceptable for predicting Ms. Thus the Scalable GPR delivers a fairly acceptable accuracy (Fig. 2(b)) with significant efficiency improvement. Using our Scalable GPR, 95% of predictions had an error of 20% or less.

Other Comparisons: Besides the conventional GPR, we also compare our method with Neural Network and SVM, which are widely used in scientific data prediction. However the Neural Network method takes more than 5 hours on training step for 870 training points and SVM delivers fairly poor predictions with low efficiency. The performances of both methods are not comparable with the Scalable GPR in terms of either efficiency or accuracy. In 2011, Słoński also showed the computational cost superiority of the GPR compared to Bayesian and standard Neural Network [7].

6   Application on Prediction of Electrical Conductivity

The efficiency and the accuracy of the Scalable GPR have been demonstrated in Ms temperature prediction. To further test the scalability of the proposed method, we conduct the second series of experiments on electrical conductivity predictions by involving a much larger scientific dataset. In this group of experiments, the conventional GPR is not able to deal with the large scale training dataset due to the extremely expensive computational cost.

6.1 Dataset

The database for Electrical Conductivity has a total of approximately 15,700 entries over 29 dimensions. This is considered to be very large as far as experimental points databases are concerned. As per the previous case study, each row has a set of chemical compositions in mol % with an associated logarithmic value of electrical conductivity in Siemens/meter. We have collected this data from the literature. In addition to the set of chemical compositions, the temperature in Celsius is also provided for each data point. The range of chemical compositions varies between 0 and 100 while the temperature varies between approximately 90 and 3,000. We rescaled the temperature by a factor of 200 to make the data more uniform and thus obtaining better predictions. In the following performance study, we will focus on the scalability our approach and discuss the effect of the batch query processing in the proposed Scalable GPR.

6.2 Performance Study

Conventional GPR: The experiments are conducted on a regular desktop computer, therefore attempting a standard GPR using a training dataset with the size of 2000×29 has proven to be very tedious and extremely slow. Thus we only randomly sample 2000 entries from the original dataset to build up the training data to test GPR. With this setting, it costs 9.28 hours for training and 5.67 minutes per prediction in the regression step. Clearly, the conventional GPR is not capable to handle real-time applications.

Scalable GPR: With the training data condensation described in Section 4, the proposed Scalable GPR can easily handle the large scale training data by capturing the intrinsic information embedded in and removing the redundant entries. We randomly select 80% entries (i.e., 12,560 entries) from the entire database to build up the initial training data set and use the remaining 20% entries as the testing points pool. Following the training data condensation described in Section 4.2, we condense the size of the training data from 12,560 points to 8,654 points by setting $\epsilon = 0.5\%$, which performs the best compared with $\epsilon = 1\%$ and 5%. We incrementally select 100, 500, 1000, 1500, 2000 and 3000 number of entries from the testing points pool as testing data to show the scalability, efficiency and accuracy of the Scalable GPR and also the effect of batch query processing on the performance. As reported in Table 3, the performance of the batch query processing is quite stable. With the size increment of the testing data from 100 to 3,000 points, the number of batch queries generated is increased from 25 to 750. The average time to create a batch query was 0.06 seconds. With different numbers of batch queries, the average training time cost, prediction (regression) time cost, and the total time cost for each testing point is very stable. With the error bound of 15%, we can always achieve the real-time prediction response averagely within 0.9 seconds.

Table 3. Scalable GPR for predicting Electrical Conductivity (target error of 15%)

Number of testing points	Number of batches	Average training time per testing point	Average prediction time per testing point	Average total time per testing point	AE (%)
100	25	0.688	0.044	0.732	14.5
500	124	0.885	0.045	0.928	14.9
1000	250	0.966	0.042	1.008	14.8
1500	375	0.741	0.043	0.784	14.2
2000	500	1.01	0.042	1.06	14.8
3000	750	0.715	0.042	0.758	14.7

7 General Discussion and Conclusion

In this paper, we propose a scalable approach to make predictions of material properties using a Gaussian process regression machine learning approach. As expected, our experiments showed that the size of the training matrix influences the calculation time exponentially. While it is clear that a very small training set would lead to poor prediction and that a large set would necessarily produce more accurate predictions, our results with Ms and Electrical Conductivity predictions show that there is no general correlation between the size of the training matrix and the predicted error when using training matrices between $10^{2.7}$ and $10^{4.7}$. We believe that the variation in prediction error is related to the quality of the data in the training matrix. In other words, closely related data in the training set will lead to better prediction. Also, since both sets of data we work with are experimental values, there is a high chance of human error in entire sets of points that could lead to variations in the results. A further analysis of the data and testing on specific systems would be required in order to create a reliable and accurate database. This would also improve the computational time as less iterations would be required to obtain the targeted error.

In summary, our approach has proven to be fast while maintaining a good prediction error. Results on prediction of Martensite Start Temperature as well as Electrical Conductivity demonstrate that the proposed Scalable GPR outperforms the other existing methods significantly in terms of efficiency and scalability.

References

1. Bailer-Jones, C., Bhadeshia, H., MacKay, D.: Gaussian process modelling of austenite formation in steel. Materials Science and Technology 15(3) (1999)
2. Gibbs, M.N., MacKay, D.J.C.: Efficient implementation of gaussian processes. Submitted to Statistics and Computing

3. Huang, Z., Shen, H., Liu, J., Zhou, X.: Effective data co-reduction for multimedia similarity search. In: Proceedings of the 2011 ACM SIGMOD International Conference on Management of Data, SIGMOD 2011, pp. 1021–1032. ACM, New York (2011)
4. Lee, S.-J., Park, K.-S.: Prediction of martensite start temperature in alloy steels with different grain sizes. Metallurgical and Materials Transactions A 44(8), 3423–3427 (2013)
5. Payson, P., Savage, C.: Martensite reactions in alloy steels. Transactions ASM 33, 261–275 (1944)
6. Rasmussen, C.E., Williams, C.K.I.: Gaussian Processes for Machine Learning (Adaptive Computation and Machine Learning). MIT Press (2005)
7. Słoński, M.: Bayesian neural networks and gaussian processes in identification of concrete properties. Computer Assisted Mechanics and Engineering Sciences 18(4), 291–302 (2011)
8. Snelson, E.: Local and global sparse gaussian process approximations. In: Proceedings of Artificial Intelligence and Statistics, AISTATS (2007)
9. Sourmail, T., Garcia-Mateo, C.: Critical assessment of models for predicting the ms temperature of steels. Computational Materials Science 34(4), 323–334 (2005)
10. Sourmail, T., Garcia-Mateo, C.: A model for predicting the ms temperatures of steels. Computational Materials Science 34(2), 213–218 (2005)
11. Stormvinter, A., Borgenstam, A., Ågren, J.: Thermodynamically based prediction of the martensite start temperature for commercial steels. Metallurgical and Materials Transactions. A 43A(10), 3870–3879 (2012), QC 20121029
12. Urtasun, R., Darrell, T.: T.: Sparse probabilistic regression for activity-independent human pose inference. In: IEEE Conference on Computer Vision and Pattern Recognition, CVPR (2008)
13. Weber, R., Schek, H.-J., Blott, S.: A quantitative analysis and performance study for similarity-search methods in high-dimensional spaces. In: Gupta, A., Shmueli, O., Widom, J. (eds.) VLDB 1998, Proceedings of 24th International Conference on Very Large Data Bases, New York City, USA, August 24-27, pp. 194–205. Morgan Kaufmann (1998)
14. de Weijer, A.P., Vermeulen, W.G., Morris, P.F., van der Zwagg, S.: Prediction of martensite start temperature using artificial neural network. Ironmaking and Steelmaking 23(5) (1996)

Mining Differential Dependencies: A Subspace Clustering Approach

Selasi Kwashie[1], Jixue Liu[1], Jiuyong Li[1], and Feiyue Ye[2]

[1] School of Information Technology & Mathematical Sciences
University of South Australia, SA 5095
[2] College of Computer Science and Engineering
Jiangsu University of Technology, Changzhou, China
selasi.kwashie@mymail.unisa.edu.au, {jixue.liu,jiuyong.li}@unisa.edu.au
yfy@jstu.edu.cn

Abstract. The discovery of differential dependencies (DDs) is the problem of finding a minimal cover set of DDs that hold in a given relation. This paper proposes a novel subspace-clustering-based approach to mine DDs that exist in a given relation. We study and reveal a link between δ-nClusters and differential functions (DFs). Based on this relationship, we adopt and co-opt techniques for mining δ-nClusters to find the set of candidate antecedent DFs of DDs efficiently, based on a user-specified distance threshold. Furthermore, we define an interestingness measure for DDs to aid the discovery of essential DDs and avoid the mining of an extremely large set. Finally, we demonstrate the scalability and efficiency of our solution through experiments on real-world benchmark datasets.

Keywords: Data dependency discovery, differential dependency, functional dependency, subspace clustering, δ-nCluster.

1 Introduction

Data dependency is a well-studied subject in the database community. The most-fundamental data dependency theory, functional dependency (FD), is a type of constraint often used to heighten data quality. The recent need for high data quality has necessitated the extension of FDs for data management applications. These extensions relax the otherwise strict equality constraint of FDs to capture different data semantics. *Differential dependency* (DD) [14] is one such extension.

A DD, denoted by $X[W_X] \to A[w_a]$, defined over a relation r states that any two tuples close on the set X of attributes ought to be close on another attribute A. The closeness is in terms of distance intervals W_X and w_a respectively. For example, given the relation in Table 1, the DD $Age[0, 2]Gen[0] \to Sal[0, 3]$ indicates that if any two employees have the same gender and their age difference is within two years, then the difference of their salaries should be within $\$3K$.

Mining data dependencies is a long-standing research problem due to the usefulness of the discoveries in real life. Over the last decades, many algorithms were developed for mining FDs and its various extensions. For instance, the works in:

H. Wang and M.A. Sharaf (Eds.): ADC 2014, LNCS 8506, pp. 50–61, 2014.

[6,11,12,16] propose methods for FD discovery; [4,5,8] present conditional FD mining techniques; [7] and [13] propose algorithms for finding soft FDs and matching dependencies respectively. The general goal of data dependency discovery is to find a *minimal cover set* of all dependencies that hold in the *entire search space* of a

Table 1. Instance of relation r

TID	Age	Gen	Edu	Sal ($K)
1	25	2	4	15
2	25	1	3	18
3	28	1	3	15
4	32	2	2	12
5	30	1	1	15

Edu: 1=Dip.; 2=BSc.; 3=M.Phil; 4=PhD.
Gen: 1=Male; 2=Female

given relation. That is, only the irreducible set of dependencies are of interest since the number of dependencies in a given relation may be extremely large.

Unfortunately, mining even a minimal cover set of DDs for a given relation is often infeasible. This is the case because, even though the task of finding DDs has the same combinatorial explosion of attribute set as other data dependencies, the search space of DDs is significantly larger as several distance functions may be defined over each attribute.

In this work, we propose the exploration of a subset of the search space of the determinant (antecedent/LHS) functions to finding DDs. Precisely, given a user-specified distance threshold, ϵ, we consider the sections of the search space where the upper-limits of the distance intervals of the LHS functions are bound by ϵ. This approach reduces the search space and increases the efficiency of our algorithm as well as produces fewer and interesting DDs. More specifically, the contributions of this paper are summarized as follows. (1) We investigate the relationship between DFs and δ-nClusters. This presents the opportunity to adopt, co-opt and utilise δ-nClusters mining algorithms to find the antecedents of DDs more efficiently. (2) Based on the derived relationship, we propose a new approach to DD discovery and present a novel algorithm for mining interesting DDs based on a user-specified distance threshold ϵ.

At the moment, the DD discovery method in [14] is the most related work to ours. The discovery algorithm in [14] is based on *reduction algorithms*: for a fixed RHS DF of each attribute in the given relation, the set of left-reduced (irreducible) LHS DFs are found to form DDs. Pruning of the search space is done based on the *subsumption* order of DFs, implication of DDs, and instance exclusion. The minimal set of DDs is then generated by the elimination of redundant DDs from the set of all left-reduced DDs.

2 Preliminaries

In this section, we introduce some preliminary concepts and definition of DDs and subspace δ-nClusters.

2.1 Differential Dependency

Let r be an instance of the relation $R(A_1, \cdots, A_n)$, $dom(A_i)$ represent the domain of an attribute $A_i \in R$, and $X, Y \subset R$ be subsets of attributes in R.

A *distance metric* d_A of an attribute A returns the distance between any two values a_1, a_2 of A in r. d_A is assumed non-negative and symmetric.

A **differential function (DF)** of an attribute A w.r.t. the distance interval $w = [x, y]^1$ returns a boolean value indicating whether $d_A(a_1, a_2)$ is within/on w or not for any two values a_1, a_2 of A. It is denoted as $A[w]$. A DF on $X = \{A_1, \cdots, A_m\}$, denoted $X[W_X]$ is: $X[W_X] = A_1[w_1] \wedge \cdots \wedge A_m[w_m]$ where $A_i \in X \wedge w_i \in \mathcal{W}_{A_i}$. $\mathcal{W}_{A_i}$ is the set of distance intervals of A_i. Given any two DFs $X[W_X]$ and $Y[W_Y]$, $X[W_X]$ is said to **subsume** $Y[W_Y]$ denoted as $X[W_X] \succeq Y[W_Y]$ iff: $\forall A_i[w_i] \in X[W_X]$, $\exists A_i[w_i'] \in Y[W_Y]$ such that $w_i' \subseteq w_i$. Let $\mathcal{T}(X[W])$ be the set of all tuple pairs that agree on (satisfy) $X[W]$.

Definition 1 (Differential Dependency (DD)). *A DD is a statement* σ : $X[W_L] \rightarrow Y[W_R]$ *between two DFs* $X[W_L], Y[W_R]$. σ *holds over* r *of* R *if and only if for any two tuples* $t_1, t_2 \in r$, *if* $X[W_L]$ *returns true,* $Y[W_R]$ *returns true.* $X[W_L], Y[W_R]$ *are termed the LHS (determinant/antecedent) and RHS (dependent/consequent) of* σ *respectively.*

Definition 2 (Minimal DD, Cover & Minimal Cover). *Let* Σ *be a set of DDs in* r. *A DD* $X[W_L] \rightarrow Y[W_R] \in \Sigma$ *is* **minimal** *if and only if the following conditions are satisfied: (a) there does not exist any DD* $X'[W_a] \rightarrow Y[W_R] \in \Sigma \wedge X'[W_a] \succeq X[W_L]$. *(b) there does not exist any DD* $X[W_L] \rightarrow Y'[W_a] \in \Sigma \wedge Y[W_R] \succeq Y'[W_a]$. *The set* Σ_1 *is a* **cover** *of the set* Σ *if each DD in* Σ *exists in or is implied by a DD in* Σ_1. *That is,* $\Sigma_1 \equiv \Sigma$. *A* **minimal cover** *set* Σ_c *is a cover set of* Σ *such that there does not exist a cover* $\Sigma' \subset \Sigma_c$.

Example 1. *Given the relation in Table 1, let* $d_A(a_1, a_2) = |a_1 - a_2|$ *for all* $A \in R$. *The DD* $Age[0] \rightarrow Edu[1]$ *holds since all tuple pairs that satisfy the DF* $Age[0]$ *also satisfy* $Edu[1]$. *If* $\Sigma = \{Age[0, 5]Gen[0] \rightarrow Edu[0]; Age[0, 5] \rightarrow Edu[0, 2]; Age[0, 7] \rightarrow Edu[0]; Gen[0]Sal[0] \rightarrow Age[3, 5]; Gen[0]Sal[0] \rightarrow Age[3, 7]\}$, *then the minimal set of* Σ *is* $\Sigma_c = \{Age[0, 7] \rightarrow Edu[0]; Gen[0]Sal[0] \rightarrow Age[3, 5]\}$.

2.2 Subspace δ-nCluster

In the following, we present the concepts and definition of a distance-based subspace clustering model, δ-nCluster, in [9].

Let $R(A_1, \cdots, A_n)$ be a feature/attribute space. The set of all tuples under R represents an instance r of R. Subspace clustering is the task of finding groups of tuples (objects) $T \subseteq r$ that are comparable/homogeneous in a subspace (set of attributes) $X \subseteq R$. Let $d_A(t_1[A], t_2[A])$ measure the distance of two tuples $t_1, t_2 \in r$ on $A \in R$. The tuple pair t_1, t_2 are said to be *neighbours* on A if $d_A(t_1[A], t_2[A]) \leq \delta \times range(A)$, where $\delta \in [0, 1]$ is a user-specified threshold; $range(A)$ is the range of distance values of A.

Given $X \subseteq R$ and a user-specified threshold δ, if t_1, t_2 are neighbours on every $A \in X$, then t_1, t_2 are said to be δ-**neighbours** of each other in the subspace X. For example, tuples 2 and 3 of the relation in Table 1 are 0-neighbours of each other in the subspace $\{Gen, Edu\}$.

---

[1] If $x = y$, $w = [x]$ for brevity.

Definition 3 (Subspace δ-nCluster). *For the pair* (T, X), *if for any pair of tuples* $t_i, t_j \in T$ *and every attribute* $A \in X$, *tuples* t_i, t_j *form a* δ-*neighbour on* A, *then the pair* (T, X) *is a subspace* δ-**nCluster**.

3 Problem Formulation

Given a relation r, the discovery of DDs is the problem of finding a minimal cover set Σ of valid DDs that hold in r. This problem inherits the inherent challenges of any dependency discovery from data, unfortunately, to a higher magnitude; due to the extremely large search space of candidate DFs. Consequently, DD discovery is not only computationally expensive, but also returns an enormously large minimal cover set, adversely affecting the utility of DDs.

In this section, we employ the semantics of DD to define an *interestingness* measure for DDs. This allows the mining of a smaller set of DDs that capture interesting patterns of data, as well as, enhances the efficiency of discovery. Next, we present the formal problem definition.

3.1 Interestingness of a DD

A DD $\sigma : X[W_X] \rightarrow A[w_a]$ models how the closeness (W_X) of a set X of attributes dictates the closeness (w_a) of a different attribute A. Hence, three intuitive factors that show the significance of σ include: (a) the closeness conveyed by W_X amongst the values of X; (b) the similarity revealed by w_a amongst the values of A; and (c) the proportion of instances that support σ.

The strength of any DD, $\sigma : X[W_X] \rightarrow A[w_a]$, is determined by its LHS DF $X[W_X]$. If W_X consists of narrow intervals, then the RHS DF, $A[w_a]$, strongly depends on X and vice versa. For instance, if for each DF $B[w_b] \in X[W_X]$, w_b covers the entire distance space of B, then $A[w_a]$ is not dependent on X. Thus, if a DD $X[W_X] \rightarrow A[w_a]$ holds, then for any $B \notin X$, $X[W_X]B[w_b] \rightarrow A[w_a]$ is implied if w_b covers all $w \in W_B$.

The RHS DF, $A[w_a]$, of σ captures the relationship amongst the values of A. Therefore, if w_a discloses a high similarity, the more useful the DD it forms, and contrariwise. For example, whereas any DD formed by the RHS DF $Age[0, 7]$ in Table 1 is trivial, those formed by $Age[0], Age[2]$ are non-trivial and potentially more useful. This is because, the DF $Age[0, 7]$ covers the entire distance space of Age, therefore, satisfied by any tuple-pair. The DFs $Age[0]$ and $Age[2]$, however, describe specific sections of the distance space, hence, do not have trivial tuple-pair satisfaction.

The support of σ represents the proportion of the relation that agree on it. We define the interestingness of a DD to capture all three factors as follows, where the functions $up(w)$ and $|w|$ on $w = [x, y]$ of a single-attribute DF $A[w]$ return y, and $|y - x|$ respectively.

Definition 4 (Interestingness of a DD). *The interestingness of a DD* σ : $X[W_X] \to A[w_a]$ *in* r *for* $X[W_X] = B_1[w_1] \cdots B_m[w_m]$ *is given by:*

$$intr(\sigma) = \frac{\Sigma_{i=1}^m prox(B_i[w_i])}{m} \times depQ(A[w_a]) \times supp(\sigma), \tag{1}$$

where $prox(B_i[w_i]) = \frac{1}{|w_i|+1}$ *reflects of the closeness requirement of LHS DF* $B_i[w_i]$; $depQ(A[w_a])^2 = \frac{|w_d|-|w_a|}{|w_d|}$ *measures the dependent quality of RHS DF* $A[w_a]$; $w_d = [u_{a_1}, u_{a_d}]$, u_{a_1}, u_{a_d} *are the minimum and maximum distance values in* $\mathcal{W}_A$ *respectively; and* $supp(\sigma) = |\mathcal{T}(X[W_X]) \cup \mathcal{T}(A[w_a])| = |\mathcal{T}(A[w_a])|$.

The definition of $intr(\sigma)$ favours DDs with fewer LHS attributes, and narrow distance interval DFs (both RHS and LHS). LHS DFs with less attributes are usually satisfied by more tuples and therefore DDs of such LHS DFs tend to have high support (hence, high $intr(\sigma)$) than DDs with more LHS attributes. However, it is noteworthy that high support is not *per se* an indicator of high interestingness. This is because although DFs with wide distance intervals usually have more support than DFs with narrow intervals, DDs of the latter DFs may be more interesting since they have higher *prox* and *depQ* values.

3.2 Problem Statement

On account of the huge search space of candidate DFs and the fact that the strength of DDs is determined by LHS DFs, we propose a user-specified distance constraint on the upper-limit y of the distance interval $w = [x, y]$ of LHS DFs. This reduces the search space of DFs while guaranteeing the discovery of interesting DDs. Specifically, the set of candidate LHS DFs is limited to those that have the upper-limit of their distance interval bound by a set threshold, ϵ. We term such DFs ϵ-DFs, defined formally as follows.

Definition 5 (An ϵ-DF). *A DF* $X[W_X] = A_1[w_1] \wedge \cdots \wedge A_m[w_m]$ *on the set* X *of attributes is said to be an* ϵ**-DF** *if for all* $A_i[w_i] \in X[W]$, $up(w_i) \leq \epsilon \times range(A_i)$, *where* $\epsilon \in [0, 1]$ *is user-specified;* $range(A_i)$ *is the range of distance values of* A_i.

Definition 6 (Problem Definition). *Given a relation* r *of* R, *a user-specified distance threshold,* ϵ, *and a minimum support value,* ms, *we find a minimal cover set* Σ_ϵ *of DDs in* r *such that* $\forall \, \sigma \in \Sigma_\epsilon, intr(\sigma)$ *is highest for the given* ms *and* $LHS(\sigma)$ *is an* ϵ-DF.

4 Relationship between ϵ-DFs and δ-nClusters

To aid the efficient mining of ϵ-DFs for DD discovery, we study and derive a link between ϵ-DFs and δ-nClusters. More importantly, this relationship enables us to co-opt and utilise δ-nCluster mining algorithms to mine LHS DFs.

---

[2] Similar to the dependent quality measure in [15].

By Definition 3, δ-nClusters exhibit the *anti-monotone property*. Thus, for any given subspace X, there may exits many δ-nClusters, unfortunately, many are redundant, in that, they are implied by others. To avoid redundancy, we define our clusters of interest, namely: δ-nClusters that are *maximal* w.r.t. tuple containment and *free* w.r.t. attribute containment.

Definition 7 (Maximal δ-nCluster, Free set of δ-nClusters). *A δ-nCluster (T, X) is* **maximal** *iff:* $\nexists (T', X)$ *such that* $T \subset T'$. *Let* Ψ_X *be the set of all maximal δ-nClusters of a subspace X. Ψ_X is* **free** *iff:* $\nexists \Psi_Y$ *such that for all* $(T, X) \in \Psi_X$, *there exist* $(T', Y) \in \Psi_Y$ *such that* $T' = T$, *where* $Y \subset X$.

The set Ψ_X of δ-nClusters in X is described as being **free-maximal** if it is both maximal (w.r.t. tuple-containment) and free (w.r.t. attribute-set-containment). Let $I_\alpha = (T_\alpha, X)$ denote a maximal δ-nCluster in X; where $\alpha \in [1, 2, \cdots, d]$ and d is the total number of maximal δ-nClusters in X.

Lemma 1 *If $\Psi_X = \{I_1, I_2, \cdots, I_d\}$ is a free-maximal set of δ-nClusters of X in r, then the ϵ-DF $X[W] = A_1[w_1] \wedge \cdots \wedge A_m[w_m]$ holds in r; where $A_1, \cdots, A_m$ are the attributes in X and $\epsilon = \delta$. Specifically,*

$$\mathcal{T}(X[W]) = pr(T_1) \cup \cdots \cup pr(T_d), \tag{2}$$

where $pr(T_\alpha)$ is the set of all tuple pairs of I_α.   □

Example 2. *Given that $\delta = 0.6$ for the relation in Table 1, let $X = \{Edu, Sal\}$. $\Psi_X = \{I_1, I_2\}$, where $I_1 = \{1, 2, 3\}$ and $I_2 = \{3, 4, 5\}$. The DF $Edu[0, 1]Sal[0, 3]$ holds with $\mathcal{T}(Edu[0, 1]Sal[0, 3]) = \{\{1, 2\}, \{1, 3\}, \{2, 3\}, \{3, 4\}, \{3, 5\}, \{4, 5\}\}$.*

Lemma 1 enables us to find ϵ-DFs and their agree tuple sets from the set of free-maximal δ-nClusters. This is the foundation for the DD discovery approach presented in the next section. Our work has shown that the mapping stated in Lemma 1 misses some DDs namely, those that do not have ϵ-DF LHSs. Because of page limit, the details of this result are omitted from this paper.

5   The Algorithm

In this section, we introduce a new DD mining approach that incorporates the δ-nCluster model in [9] from data mining into the context of DD discovery. The pseudo-code of our algorithm is presented in Algorithm 1[3]. It consists of four major steps namely: finding the set of all free-maximal δ-nClusters; forming candidate LHS DFs; finding valid RHS DFs and forming DDs; and pruning of derived set of DDs. In the following subsections, we elaborate each step.

5.1   Finding Free-Maximal δ-nClusters

This step finds all free-maximal δ-nClusters in the given relation, r, using Algorithm 2. It is an adaptation of the δ-nClusters mining framework in [10].

---

[3] SCAMDD: Subspace Clustering-based Approach to Mining DD.

In lines 2–3 of Algorithm 2, we employ the approach used in [9,10] to find the maximal δ-nClusters of single attributes. For each $A \in R$, let r_A be the sorted (in an ascending order) instance values of A in r. Let p_1, p_2 be two positions in r_A such that $p_1 < p_2$. The set T_{12} of tuples (from p_1 to p_2) form a maximal δ-nCluster in A if and only if: (a) $d_A(p_1[r_A], p_2[r_A]) \leq \theta$; (b) $d_A((p_1 - 1)[r_A], p_2[r_A]) > \theta \wedge d_A(p_1[r_A], (p_2 + 1)[r_A]) > \theta$, where $\theta = \delta \times range(A) \wedge \delta = \epsilon$.

Next, a database $\mathcal{D}$, of *attribute-lists* is generated from the single-attribute maximal δ-nClusters (line 4). An attribute-list of a tuple t is the list of clusters in which t has no less than $(mr - 1)$ δ-neighbours. The minimum number of tuples mr that are allowed to form a cluster in our case is two. The multiple-attribute δ-nClusters are then mined from the collection of all attribute-lists as frequent patterns (line 5) using an efficient implementation of the frequent pattern mining algorithm, $FPGrowth$ in [3]. In line 6, a hash-table $\mathcal{C} = \langle X, \Psi_X \rangle$, of all free-maximal δ-nClusters is created, where $X \subset R$ and Ψ_X is the set of free-maximal δ-nClusters in X.

Algorithm 1. $SCAMDD(r, \epsilon, ms)$

Input: The relation r of R, user-specified threshold ϵ, minimum support, ms.

Output: A minimal cover set Σ_ϵ of DDs.

1: Find all free-maximal δ-nClusters in r
2: Form candidate LHS DFs
3: Find valid RHSs for candidate LHSs and form DDs
4: Prune the set Σ of DDs
5: **return** Σ_ϵ (Pruned Σ)

Algorithm 2. $freeMax(r, \delta)$

Input: $r, \delta = \epsilon$.

Output: The set $\mathcal{C}$ of Ψ_X δ-nClusters

1: **function** FREEMAX(r, δ)
2:     **for each** $A \in R$ **do**
3:         find all maximal δ-nClusters in A
4:         create database $\mathcal{D}$ of attribute-lists
5:         mine all free-maximal δ-nClusters in $\mathcal{D}$
6:         create a Hash-table $\mathcal{C} = \langle X, \Psi_X \rangle$
7:     **return** $\mathcal{C}$

5.2 Forming Candidate LHS DFs

Here, the set of candidate LHS DFs are derived from the sets of free-maximal δ-nClusters, $\mathcal{C}$, in Algorithm 3. We use the attribute lattice structure [1] to generate the set of candidate LHSs as follows. Each node N in the lattice is a triplet $N = (e, f, g)$ where $e = X$, a subset of R; $f = X[W]$, an ϵ-DF of X; and $g = \mathcal{T}(X[W])$, the list of all tuple pairs that satisfy $X[W]$.

At the first level (lines 2–6), for every $A \in R$, given that there exists the set Ψ_A of free-maximal δ-nClusters of A in $\mathcal{C}$ (line 4), then $A[w_a]^4$ and $\mathcal{T}(A[w_a])$ are generated according

---

Algorithm 3. formCandLHSs$(\mathcal{C})$

Input: $\mathcal{C}$

Output: Lattice $\mathcal{L}$ of candidate LHS DFs

1: **for** Level-$i = 1 \rightarrow n - 1$ **do**
2:     **if** $(i == 1)$ **then**
3:         **for each** attribute $A \in R$ **do**
4:             **if** $(\exists \Psi_A \in \mathcal{C})$ **then**
5:                 create node, $N_A(A, A[w], T(A[w]))$ by Lemma 1 add N_A to $\mathcal{L}_1$
6:             add $\mathcal{L}_1$ to $\mathcal{L}$
7:     **else**
8:         **for** $l = 1, \cdots, |\mathcal{L}_{i-1}|$ **do**
9:             **for** $k = l + 1, \cdots, |\mathcal{L}_{i-1}|$ **do**
10:                **if** $(|(N_l).e \cap (N_k).e| == (i - 2) \wedge (\exists \Psi_X \in \mathcal{C}))$ **then**
11:                    form $\mathcal{L}_i$ node with $N_l, N_k \in \mathcal{L}_{i-1}$
12:                add $\mathcal{L}_i$ to $\mathcal{L}$
13: **return** $\mathcal{L}$

---

^{4} $w_a = [x, y]$ where x, y are the min. and max. distance of values in Ψ_A respectively.

to Lemma 1 to form the node $N_A(A, A[w_a], \mathcal{T}(A[w_a]))$ (line 5). The nodes at level–1 form the first level, $\mathcal{L}_1$, of the lattice $\mathcal{L}$ (line 6).

Nodes of any level-i ($2 \le i < n$), $\mathcal{L}_i$, are formed from $\mathcal{L}_{i-1}$ nodes (lines 7–12). Any two nodes $N_l, N_k \in \mathcal{L}_{i-1}$ are parent to a node N_c of $\mathcal{L}_i$ iff: N_l, N_k have up to $(i-2)$ preceding single-attributes in common on their first-triplet e (set of attribute, sorted in lexicographical order) and their remaining attributes in e are different (condition 1 of line 10). If this condition is satisfied, then $(N_c).e = X = \{(N_l).e\} \cup \{(N_k).e\}$. If there exists $\Psi_X \in \mathcal{C}$ (condition 2 of line 10), then node $N_c(X, X[W_X], \mathcal{T}(X[W]))$ of $\mathcal{L}_i$ is formed (by Lemma 1) with $N_l, N_k \in \mathcal{L}_{i-1}$ as parents (line 11).

For example, the lattice of a relation with 4 attributes is show in Fig. 1, assuming there exist free-maximal δ-nCllusters for all $X \subset R$. Each attribute forms a node in the first level ($\mathcal{L}_1$). $\mathcal{L}_2$ nodes are combination of all $\mathcal{L}_1$ nodes; at $\mathcal{L}_3$, the $\mathcal{L}_2$ pair AB, AC for instance, can form a node $N.e = ABC$. The lattice ends at $\mathcal{L}_3$ since

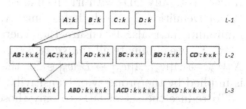

Fig. 1. The lattice of a 4 attribute relation

$n = 4$. Unlike the FD lattice, at each node N, assuming each $A \in R$ has d distinct distance values, there exists k^m possible DFs; where $k = d! = \frac{d(d+1)}{2} \wedge m = |\{(N).e\}|$.

5.3   Finding Valid RHS DFs to form DDs

Given a LHS DF $X[W_L]$, a candidate RHS $A[w_a]$ is *valid* iff: (1) $X \cap \{A\} = \emptyset$; (2) $\mathcal{T}(X[W_L]) \subseteq \mathcal{T}(A[w_a])$; (3) $\nexists\, A[w'_a]$ such that $A[w_a] \succeq A[w'_a] \wedge \mathcal{T}(A[w'_a]) \subseteq \mathcal{T}(X[W_L])$. The first requirement ensures that $A[w_a]$ has only one attribute A, and $A \notin X$ to avoid the generation of redundant and implied DDs. Conditions 2 and 3 require that $A[w_a]$ forms a valid DD with $X[W_L]$ in r (by Definition 1) and no implied DD is found respectively.

The function *findRHSs* in Algo-

Algorithm 4. findRHSs($\mathcal{L}$)
1: **function** FINDRHSs($\mathcal{L}$)
2:    **for** $\mathcal{L}_i \in \mathcal{L}; i = 1 \to n$ **do**
3:       **for each** node $N \in \mathcal{L}_i$ **do**
4:          **if** $N.f$ is valid **then**
5:             $\mathcal{R} = \{\}$ and $Y = \{R/X\}$
6:             **for each** $A \in Y$ **do**
7:                calculate w_a using $(N.g, A)$
8:                **if** $depQ(A[w_a]) \neq 0$ **then**
9:                   add $A[w_a]$ to $\mathcal{R}$
10:            **if** $\mathcal{R} \neq \emptyset$ **then**
11:               **for each** $A[w_a] \in \mathcal{R}$ **do**
12:                  form DD $\sigma : N.f \to A[w_a]$
13:                  Add σ to Σ
14:    **return** Σ

rithm 4 captures the search strategy for the set of valid RHSs for all candidate LHS DFs in $\mathcal{L}$. For each level i ($1 \le i \le n$) in $\mathcal{L}$, for every node $N \in \mathcal{L}_i$, if there exists no DD amongst the ϵ-DF, $(N).f$, of N, (line 4), then we find the set $\mathcal{R}$ of valid RHS DFs for $(N).f$ (line 5–9). Let Y be the set of candidate attributes to form a valid RHS DF with $(N).f$. $Y = \{R/X\}$, where $X = (N).e$. For each $A \in Y$, the projection of values of A with respect to the set of agree tuples $(N).g$ of the node N is generated (line 7). The minimum and maximum distance values in the projection forms the distance interval w_a of A. If w_a does not cover the distance space of A (line 8), it is added to the set $\mathcal{R}$ (line 9). If valid RHSs are found (line 10), they form DDs with $(N).f$ (lines 11–13).

5.4 Pruning

The final stage of the discovery process is the generation of the minimal set Σ_ϵ of DDs from the set Σ of valid DDs. We discus in the following pruning operations on the set of valid DDs Σ to ensure Algorithm 1 returns Σ_ϵ according to Definition 2. The first pruning operation is based on condition (a) of Definition 2. The set Σ of valid DDs is partitioned into subsets of DDs with common RHSs. In each set, only DDs with irreducible LHSs are added to the set Σ_ϵ. Condition (b) of Definition 2 already holds since Algorithm 4 finds RHSs with the most minimum intervals. To ensure that there exists no cover Σ_1 of Σ such that $\Sigma_1 \subset \Sigma_\epsilon$, we eliminate all *transitively-implied* DDs. That is, given that the DDs $X[W_x] \rightarrow A[w_a], A[w_a] \rightarrow B[w_b] \in \Sigma_\epsilon$, then any DD of the form $X[w_x] \rightarrow B[w_b]$ is implied by the property of transitivity, hence removed from Σ_ϵ.

To reduce the set of DDs to only those that capture essential knowledge in data, we utilise the support and interestingness measures to prune candidate LHSs DFs and DDs respectively. Given a minimum support ms, all candidate LHS DFs with support less than ms are pruned out. The reason is that, if $support(X[W_X]) < ms$, then all DDs with $X[W_X]$ as LHS DF have support less than ms since $support(X[W_X] \cup A[w_a]) < ms$ for any RHS DF $A[w_a]$. Furthermore, for a specified minimum interestingness value $min\text{-}intr$, we prune any DD σ with $intr(\sigma) < min\text{-}intr$. Where no $min\text{-}intr$ is given, we mine DDs with highest interestingness for the given ms.

6 Experimental Evaluations

All the proposed algorithms in this paper are implemented in Java. The experiments were conducted on an Intel Core i5-2520M CPU @ 2.5GHz processor computer with 4.0 GB of memory running Windows 8 OS. The data sets used in the experiments are briefly described in Table 2, downloaded from the UCL Machine Learning data repository [2].

Table 2. Description of data sets

Data sets	Size	No. of Attributes
Adult	26690	15 (6 N; 9 C)
Red-wine	1599	12 (12 N; 0 C)
Breast Cancer	569	31 (30 N; 1 C)
Thoracic	470	17 (3 N; 14 C)
Iris	150	5 (4 N; 1 C)

N = numeric, C = categorical

For any data set, various distance functions can be defined for each attribute based on domain-knowledge. For our experiments, we use the absolute values of difference as distance metric for all numeric attributes. For categoric attributes, we use the equality function.

6.1 Time Performance and Scalability

We conduct experiments to show: (a) the practical infeasibility of mining DDs in the entire search space of all possible DFs in relation; (b) how our algorithm

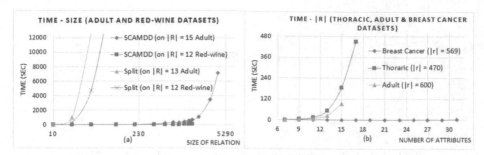

Fig. 2. Time performance on varying tuples and attribute sizes

performs w.r.t. varying number of tuples *and* attributes of a relation. The results are displayed in Fig. 2.

To show point (a) above, we implement the proposed algorithm in [14] (**Split[5] algorithm**), which does *not* require constraint on LHS DFs. In this experiment, whereas we allow the Split algorithm to mine DDs in the entire search space of DFs for various instance sizes of the Adult and Red-wine datasets, we employ the SCAMDD algorithm to find DDs with the strictest constraint on LHS DFs (set $\epsilon = 0$). Figure 2 (a) shows the plot of the time performance (on y-axis) of the two algorithms against various instance sizes (on x-axis) of the two datasets. From the graphs, it is clear that, finding DDs in the entire search space quickly becomes infeasible with increasing tuple sizes. We note that, the plot is not a comparison of time performance of the two methods (since they find different DDs) but a demonstration of the practical infeasibility of mining DDs in the entire search space of DFs. Furthermore, Fig. 2 (a) shows how the SCAMDD algorithm responds to increasing relation size of different datasets (for $\epsilon = 0$). SCAMDD performs generally well.

In the next set of experiments, we show how the SCAMDD algorithm performs on increasing attribute size of relations. For this experiment, we used the whole instances of the Breast Cancer and Thoracic datasets and 600 tuples of the Adult dataset for fair comparison of results among the datasets. Figure 2 (b) displays the results. From the plots of the three datasets, it is clear that the number of attributes alone *per se* does not determine the time performance. A look Table 2 reveals that datasets with more categorical attributes have higher time performance. This is the case, mainly because the set of instance values of categorical attributes is usually less than that of numeric attributes. Therefore, the size of tuple sets that form any cluster is generally larger, hence the higher execution time for finding associated ϵ-DFs and DDs.

6.2   Effect of the Parameters

SCAMDD uses two parameters namely, user-specified distance threshold, ϵ, and minimum support, ms. The next category of experiments examine the effect of these parameters.

---

[5] we name it Split since it 'splits' the search space.

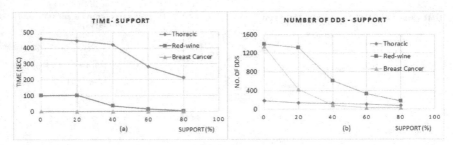

Fig. 3. Effect of support on time performance and number of DDs

As discussed earlier in subsection 5.4, ms and $min\text{-}intr$ are used to prune candidate LHS DFs and DDs respectively. The effect of ms on time performance and number of discovered DDs is shown in Fig. 3 (a) and (b) respectively. As expected, time performance improves with increasing support as more candidate LHS DFs get pruned, leading to fewer DDs that represent persistent patterns in the datasets.

For a given ϵ, only DDs with their LHS DF $X[W]$ having $prox(X[W]) \geq 1/wid([0,\epsilon])$ are of interest. Furthermore, since DFs with highest $depQ$ values are found for all LHS DFs for any ms value, it is clear that SCAMDD returns only the most interesting DDs. For all experiments, ϵ is set to minimum (0). Thus, the $prox\,(intr(\sigma))$ values for all candidate LHS DFs (σ) are maximum for any ms.

Increasing ϵ relaxes the constraint on the LHS DFs. This increases the search space of DFs and significantly affects the time performance adversely. Figure 4 shows how time performance (on y-axis) quickly increases w.r.t. increasing ϵ (on x-axis) for even the least dataset in Table 2, Iris. From this demonstration, we note that SCAMDD is sensitive to high ϵ.

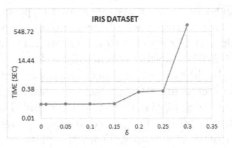

Fig. 4. Effect of relaxing ϵ on time

7   Conclusion

This paper proposes an algorithm, SCAMDD, for mining DDs based on a user-specified distance threshold, ϵ. SCAMDD uses techniques similar to finding δ-nClusters to discover candidate LHS DFs of DDs. Results from experiments show that SCAMDD is very efficient when constraint on LHS DFs is at its strictest.

The proposition is sensitive to relaxed constraints on LHS DFs due to the gigantic nature of the search space of possible LHS DFs. In our next studies, we will investigate further pruning strategies to enable efficient discovery of DDs for relaxed constraints (high ϵ values) as well as the utilization of DDs in data management application fields like data quality repair.

Acknowledgement. This work was partially supported by Australian Research Council Discovery grant DP130104090.

References

1. Agrawal, R., Srikant, R.: Fast Algorithms for Mining Association Rules in Large Databases. In: 20th International Conference on Very Large Data Bases, pp. 487–499. Morgan Kaufmann Publishers Inc. (1994)
2. Bache, K., Lichman, M.: UCI Machine Learning Repository (2013), http://archive.ics.uci.edu/ml
3. Borgelt, C.: Frequent Item Set Mining. Wiley Interdisciplinary Reviews: Data Mining and Knowledge Discovery 2(6), 437–456 (2012)
4. Fan, W., Geerts, F., Li, J., Xiong, M.: Discovering Conditional Functional Dependencies. IEEE Trans. on Knowledge and Data Engineering 23, 683–698 (2011)
5. Golab, L., Karloff, H., Korn, F., Srivastava, D., Yu, B.: On Generating Near-optimal Tableaux for Conditional Functional Dependencies. Proc. VLDB Endow. 1(1), 376–390 (2008)
6. Huhtala, Y., Krkkinen, J., Porkka, P., Toivonen, H.: Tane: An Efficient Algorithm for Discovering Functional and Approximate Dependencies. The Computer Journal 42(2), 100–111 (1999)
7. Ilyas, I.F., Markl, V., Haas, P.J., Brown, P., Aboulnaga, A.: CORDS: Automatic Discovery of Correlations and Soft Functional Dependencies. In: International Conference on Management of Data, pp. 647–658 (2004)
8. Li, J., Liu, J., Toivonen, H., Yong, J.: Effective Pruning for the Discovery of Conditional Functional Dependencies. Computer Journal 56(3), 378–392 (2013)
9. Liu, G., Li, J., Sim, K., Wong, L.: Distance Based Subspace Clustering with Flexible Dimension Partitioning. In: 23rd International Conference on Data Engineering, pp. 1250–1254 (2007)
10. Liu, G., Li, J., Sim, K., Wong, L.: Efficient Mining of Distance-based Subspace Clusters. Statistical Analysis and Data Mining 2(5-6), 427–444 (2009)
11. Liu, J., Ye, F., Li, J., Wang, J.: On Discovery of Functional Dependencies from Data. Data & Knowledge Engineering 86, 146–159 (2013)
12. Novelli, N., Cicchetti, R.: FUN: An Efficient Algorithm for Mining Functional and Embedded Dependencies. In: Van den Bussche, J., Vianu, V. (eds.) ICDT 2001. LNCS, vol. 1973, pp. 189–203. Springer, Heidelberg (2000)
13. Song, S., Chen, L.: Discovering Matching Dependencies. In: 18th ACM Conference on Information and Knowledge Management, pp. 1421–1424 (2009)
14. Song, S., Chen, L.: Differential Dependencies: Reasoning and Discovery. ACM Trans. Database Syst. 16, 1–16 (2011)
15. Song, S., Chen, L., Cheng, H.: Parameter-Free Determination of Distance Thresholds for Metric Distance Constraints. In: 28th International Conference on Data Engineering, pp. 846–857. IEEE Computer Society (2012)
16. Wyss, C., Giannella, C., Robertson, E.: FastFDs: A Heuristic-Driven, Depth-First Algorithm for Mining Functional Dependencies from Relation Instances Extended Abstract. In: Kambayashi, Y., Winiwarter, W., Arikawa, M. (eds.) DaWaK 2001. LNCS, vol. 2114, pp. 101–110. Springer, Heidelberg (2001)

A Study on the Applications of Emerging Sequential Patterns

Vincent Mwintieru Nofong[*], Jixue Liu, and Jiuyong Li

School of Information Technology and Mathematical Science,
University of South Australia
vincent.nofong@mymail.unisa.edu.au,
{Jixue.Liu,Jiuyong.Li}@unisa.edu.au

Abstract. This article presents a study on the techniques for detecting Emerging Sequential Patterns (ESPs) and the effectiveness of predictions made by ESPs in time-stamped datasets. ESPs are sequential patterns whose frequencies increase from one time-stamp dataset to another. ESPs capture emerging trends with time in sequential datasets and they are proposed for trend prediction. This work presents a study on the effectiveness of such predictions made by ESPs. Our experimental results show that, ESPs improve patterns' re-occurrence prediction than frequent patterns, but the improvements are marginal. Further more, we note that both ESPs and frequent patterns do not fare well in predicting the continuous emergence of patterns with time. Hence, we conclude with suggestions on future works that will improve current ESPs definition to enable detect non-trivial and interesting ESPs which can help increase the precision of predicting future emerging patterns with ESPs.

Keywords: Data mining, Sequential Patterns, Emerging Patterns, Pattern matching, Trend Prediction, Decision making.

1 Introduction

Emerging pattern mining is a well-studied subject in the data mining community. Emerging Patterns (EPs), are defined as patterns whose frequencies change significantly from one dataset to another. EPs in static datasets such as those with classes (male vs. female, cured vs. not cured), reveal useful and hidden contrast patterns between datasets for various decision making, for instance, in constructing accurate classifiers [7,13,14], in disease likelihood prediction [12], discovering patterns in gene expression data [15], and so on.

Emerging Sequential Patterns (ESPs), are sequential patterns whose frequencies increase with time from one sequential dataset to another (in this work *time-stamped datasets* and *sequential datasets* are used interchangeably). ESPs capture emerging trends with time in sequential datasets that can be easily understood and used in various decisions making. For example, ESPs in sequential transactions of a shop, indicating the emerging trends in customers' interest can

---

[*] Corresponding author.

H. Wang and M.A. Sharaf (Eds.): ADC 2014, LNCS 8506, pp. 62–73, 2014.

be exploited by management to understand customers' behaviours [10,22,23] and predict future purchases.

Mining emerging sequential patterns to the best of our knowledge, was introduced recently by Tsai et al in [23]. They proposed a framework for detecting changes and ESPs in customers' behaviour (transactions) mined as crisp (general) sequential patterns. Their approach was later adapted in [22] to detect changes and ESPs in customers' transactions mined as time-interval sequential patterns. Huang et. al in [10] also modified the technique proposed in [23] to detect changes and ESPs in customers' transactions mined as fuzzy time-interval sequential patterns. ESPs detected by these works provide valuable references to retail management in understanding emerging trends in customers' behaviour (purchases), and for predicting future customer behaviours.

This work presents a study on the techniques for detecting ESPs in time-stamped datasets and the effectiveness of predictions made by detected ESPs. The objective of this study is to assess ESPs based on current definition through an evaluation on the effectiveness of predictions made by ESPs versus those made by frequent patterns.

This study contributes to research by providing an in-depth understanding of the definition and applications of ESPs in time-stamped datasets. Additionally, it outlines future works based on experimental results that can help users and researchers discover non-trivial and useful ESPs for effective decision making in time-stamped datasets.

The rest of the paper is organized as follows. The basic concepts of sequential pattern mining in time-stamped datasets are presented in Section 2. A review of the definition for ESPs in sequential datasets, the processes for detecting ESPs and their usefulness are presented in Section 3. Based on the implications of ESPs in time-stamped datasets, Section 4 presents an experimental evaluation on the effectiveness of predictions made by ESPs versus that of frequent patterns. From the experimental results, Section 5 presents conclusions, recommendations, and an outline of future works on detecting non-trivial and more useful emerging sequential patterns in time-stamped datasets.

2   Basic Concepts

2.1   Sequential Patterns

The problem of mining sequential patterns and its associated notation can be given as follows.

Let I = $\langle i_1, i_2,..., i_n \rangle$ be a set of literals, termed items, which comprise the alphabet. An event is a nonempty unordered collection of items. It is assumed without loss of generality that items of an event are sorted in lexicographic order. A sequence is an ordered list of events. An event is denoted as $\langle i_1, i_2,..., i_k \rangle$, where i_j is an item. A sequence S is denoted as $\langle (a_1), (a_2), (a_3), ..., (a_n) \rangle$, where a_i is an event. A sequence with k-items, where $k = \sum_j |a_j|$, is termed a k-sequence. For example, $\langle a, b, c \rangle$ is a 3-sequence. A sequence $S_1 = \langle b_1, b_2, ..., b_k \rangle$

is a *subsequence* of $S_2 = \langle \beta_1, \beta_2, ..., \beta_n \rangle$ and S_2 is a *super-sequence* of S_1 if there exist integers $1 \leq i_1 < i_2, ..., < i_k \leq n$ such that $b_1 \subseteq \beta_{i_1}, b_2 \subseteq \beta_{i_2}, ..., b_n \subseteq \beta_{i_n}$.

Given a database D of input sequences where each input sequence has the fields: sequence-id, event-time, and the items of events. Assuming that no sequence has more than one event with the same time-stamp, the time-stamp can be used as the event identifier. The *support* of a sequence S, denoted $sup(S)$, is defined as the number of input sequences in the database D of which S is a subsequence.

Given a set of data sequences, sequential pattern mining is the process of discovering all subsequences that are frequent, that is, finding all subsequences in the database whose supports exceeds a user specified minimum support [8].

3   Emerging Sequential Patterns

3.1   Definition

Given two time-stamped datasets, D^t and D^{t+1}. If S^t and S^{t+1} are sets containing sequential patterns mined from D^t and D^{t+1} respectively, such that patterns $s_i^t \in S^t$ and $s_j^{t+1} \in S^{t+1}$, for $i = 1, 2, 3, ... |S^t|$; $j = 1, 2, 3, ... |S^{t+1}|$:

Definition 1. [ESP [10,22,23]] *A sequential pattern $s_j \in D^{t+1}$ is an emerging sequential pattern if for a given minimum support:*

1. *s_j occurs as a sequential pattern in both D^t and D^{t+1}, and*
2. $\frac{sup(s_j^{t+1}) - sup(s_j^t)}{sup(s_j^t)} > \delta$.

where δ is a user defined minimum support increment.

3.2   Discovering Emerging Sequential Patterns

Given a number of time-stamped datasets, $D^{t(i)}, D^{t(i+1)}, D^{t(i+2)}, D^{t(i+3)} ..., D^{t(i+n)} \mid i > 0$, to discover ESPs in any two consecutive time-stamped datasets, the following three processes are involved:

1. Sequential patterns are mined with a specified minimum support from the two consecutive time-stamped datasets for example, $D^{t(i)}$ and $D^{t(i+1)} \mid i > 0$.
2. All patterns in $D^{t(i)}$ are compared with the patterns in $D^{t(i+1)}$ to identify their respective identical patterns in $D^{t(i+1)}$ (that is, satisfying condition 1 of Definition 1).
3. For any two identical patterns, for example, $s_a \in D^{t(i)}$ and $s_b \in D^{t(i+1)}$, condition 2 of Definition 1 is tested. If the support change between s_a and s_b is greater than δ, the pattern $s_b \in D^{t(i+1)}$ is classified as an emerging sequential pattern of $s_a \in D^{t(i)}$.

3.3   Usefulness of ESPs

ESPs are very useful in various decision making, for example, given a set of emerging sequential patterns $E_{t(i)}^{t(i+1)}$, detected from $D^{t(i)}$ to $D^{t(i+1)}$:

1. A prediction of a sequential pattern $s_a \in D^{t(i+2)}$ can be made from the inference that $s_a \in E_{t(i)}^{t(i+1)}$.

2. A prediction of the set of emerging sequential patterns $E_{t(i+1)}^{t(i+2)}$ from $D^{t(i+1)}$ to $D^{t(i+2)}$ can also be made.

For any two time-stamped datasets, emerging sequential patterns provide quick and valuable references about the datasets which can be used in various decision making. For example, in time-stamped datasets of customers' transactions, ESPs in customers' behaviour have been employed in [5,6,10,20,22,23] to understand and/or predict customer behaviours with time.

In this study, we assess the effectiveness of predictions made by ESPs (based on current definition) versus same predictions made with frequent patterns on real-world datasets. To reduce the number of redundant patterns in our experiments, we mine sequential patterns as:

1. Closed sequential patterns: A closed sequential pattern S_n, is a frequent sequence mined from a sequence database such that there exist no super-pattern S_m with the same support as S_n. The technique proposed in [9] is employed in this study to mine this pattern variant for our experiments.

2. Closed and Maximal sequential patterns: A closed and maximal sequential pattern S_i, is a frequent sequence mined from a sequence database such that there exist no super-pattern S_j that is either frequent or have the same support as S_i. We employ the technique proposed in [21] to mine this pattern variant for our experiments.

4   Empirical Assessments

4.1   Datasets

The following datasets were used in the study:

1. **Sales_Fact_1997 & Sales_Fact_1998 Datasets:** These datasets were obtained from the $FoodMart2000$ transaction database provided by Microsoft SQL Server 2000. Sales_Fact_1997 contains 20522 transactions from 5581 customers in year 1997 and Sales_Fact_1998 contains 34015 transactions from 7824 customers in year 1998. Each dataset has 1559 unique items with 110 product classes and 47 product categories. Eight consecutive seasons are available from both datasets combined. To avoid trivial mining results, product items were aggregated to product categories for all transactions.

2. **TaFeng Retail Dataset.**
   This dataset obtained from AIIA Lab (http://aiia.iis.sinica.edu.tw) comprises of customer transactions from TaFeng Warehouse. It contains 23812 unique items, 817741 transactions from 32266 unique customers within a four month period (Nov & Dec 2000, and Jan & Feb 2001). Only four consecutive time-periods are available from this dataset.

3. **Belgium Retail Dataset.**
   This is a retail basket data of by an anonymous Belgium retail supermarket obtained from FIMI repository (http://fimi.ua.ac.be). Collected over three non-consecutive periods, (first half of December 1999 to first half January 2000, beginning of May 2000 to the beginning of June 2000, and from the end of August 2000 to the end of November 2000). No data exists between these periods. 5133 unique customers purchased at least one product from the supermarket within the 5 months period of data collection amounting to 88162 total transactions. This dataset was divided into eight time-stamped datasets.

4.2   Experimental Design

To assess the effectiveness of predictions made by ESPs (detected based on current definition) in time-stamped datasets, four main objectives, listed as the following were investigated:

1. **Objective 1:**
   - How effective are ESPs detected from two consecutive time-stamped datasets in predicting the frequent patterns of the next immediate time-stamped dataset?

2. **Objective 2:**
   - How effective are ESPs from two consecutive time-stamped datasets $D^{t(i)}$ to $D^{t(i+1)} \mid i>0$, in predicting the emerging sequential patterns from $D^{t(i+1)}$ to $D^{t(i+2)}$?

3. **Objective 3:**
   - How effective are ESPs from two consecutive seasons in predicting the frequent patterns of the next recurring seasons?

4. **Objective 4:**
   - How effective are ESPs from two consecutive seasons in predicting the ESPs of the next recurring seasons?

These objectives were investigated by carrying out the following experimental steps:

1. **Objective 1:**
   (a) For any three consecutive time-stamped datasets; $D^{t(i)}$, $D^{t(i+1)}$ and $D^{t(i+2)} \mid i>0$, sequential patterns are mined with a minimum support.
   (b) The set of ESPs in sequential patterns from $D^{t(i)}$ to $D^{t(i+1)}$ are detected.
   (c) With the detected ESPs, the frequent patterns in $D^{t(i+2)}$ are predicted [predicted patterns = set of detected ESPs].

(d) The predicted patterns, are then compared with the actual patterns in $D^{t(i+2)}$.

(e) The precision of pattern prediction by ESPs, Pr_1, is evaluated [Table 1 shows the precision evaluation formulas].

(f) With the frequent patterns mined from $D^{t(i+1)}$, the frequent patterns in $D^{t(i+2)}$ are predicted and precision of pattern prediction by frequent patterns, Pr_2, evaluated.

(g) Steps (a), (b), (c), (d), (e) and (f) are repeated on the same datasets with different sets of minimum supports: $\{0.1, 0.01, 0.001\}$, $\{0.1, 0.08, 0.064\}$ and $\{0.1, 0.04, 0.008\}$. The process is repeated for all sets of three consecutive time-stamped datasets in each dataset.

2. **Objective 2:**

(a) Same processes as in Objective 1, however instead of predicting frequent patterns, emerging patterns are predicted in steps (c), and (f). Precisions Pr_3, and Pr_4 are evaluated in steps (e) and (f) for the predictions made by ESPs and frequent patterns respectively.

3. **Objective 3:**

(a) For any two consecutive seasons (e.g. 1st winter and 1st spring) and their recurring seasons (2nd winter and 2nd spring) sequential patterns are mined from each dataset with a given minimum support.

(b) The set of ESPs in the first consecutive seasons are detected. With the ESPs, the frequent patterns in the next recurring seasons are predicted.

(c) The predicted patterns are compared with the actual patterns at the recurred seasons to evaluate their prediction precisions $Pr_{1,1}$ and $Pr_{1,2}$ respectively.

(d) With frequent patterns mined from the first seasons, the frequent patterns in the next recurring seasons are predicted and the prediction precision $Pr_{2,2}$ evaluated.

(e) Steps (a), (b), (c) and (d) are repeated with different sets of minimum supports: $\{0.1, 0.01, 0.001\}$, $\{0.1, 0.08, 0.064\}$ and $\{0.1, 0.04, 0.008\}$. The process is repeated for all two consecutive seasons and their recurring seasons.

4. **Objective 4:**

(a) Same processes as in Objective 3, however instead predicting frequent patterns, emerging patterns are predicted in steps (b) and (d). Precisions $Pr_{e,1}$ and $Pr_{e,2}$ are evaluated in steps (c) and (f) for the predictions made by ESPs and frequent patterns respectively.

4.3   Results and Discussions

Results and Discussion for Objectives 1 and 2. Tables 2 and 3 show the experimental results on one set of minimum supports $\{0.1, 0.04, 0.008\}$ with $\delta = 0$ for objectives 1 and 2 on the Sales Fact and TaFeng retail datasets respectively. Similar results were obtained for the other two sets of minimum supports ($\{0.1, 0.01, 0.001\}$ and $\{0.1, 0.08, 0.064\}$) with varying values for δ.

1. **Objective 1**
   The prediction precisions made by ESPs are slightly higher than that by frequent patterns but the increase is marginal, on the average, only +2.66% for the Sales Fact dataset and +3.45% for the TaFeng dataset as shown in columns Δ_1 of Tables 2 and 3 respectively. ESPs thus do not make significantly better predictions of patterns' re-occurrence in time-stamped datasets compared to same predictions made with frequent patterns.
2. **Objective 2**
   The prediction precisions made by ESPs are lower than that by frequent patterns by marginal averages of -8.05% in the Sales Fact dataset and -1.48% in the TaFeng dataset as shown in columns Δ_2 of Tables 2 and 3 respectively. ESPs thus make slightly lower precise predictions of emerging patterns in time-stamped datasets compared to the precisions of randomly guessing emerging patterns from the set of frequent patterns.

Results and Discussions for Objectives 3 and 4. The results of these experiments are shown in Table 4 for Sales Fact dataset with one set of minimum support $\{0.1, 0.04, 0.008\}$ and $\delta = 0$. Similar results were obtained for the other two sets of minimum supports with varying values for δ.

1. **Objective 3**
   The prediction precisions made by ESPs are again slightly higher than that of frequent patterns with a marginal average of only +0.31% as shown in column $\Delta_{1,2}$ of Table 4. Hence ESPs do not make significantly better predictions for the cyclic or seasonal re-occurrence of patterns than randomly guessing their re-occurrence with frequent patterns.
2. **Objective 4** The prediction precisions made by ESPs are lower than that by frequent patterns by a marginal average of -3.97% as shown in column Δ_e of Table 4. ESPs thus make slightly lower precise predictions of emerging patterns in the next recurring seasons compared to the precisions of randomly guessing emerging patterns in the next recurring season from the set of frequent patterns.

4.4   General Conclusions

From the study, the following general conclusions were made:

1. For predicting the occurrence of patterns in the following time-stamp, ESPs make high precise prediction. However, the precisions are not significantly higher than those made by frequent patterns. In other words, frequent patterns can make nearly as good pattern predictions as ESPs.
2. For predicting the continuous emergence of a pattern in the following time-stamp, ESPs make very low precise precisions. Randomly guessing emerging patterns in the following time-stamp with frequent patterns in the previous time-stamp results in slightly higher precise predictions than those made by ESPs. Hence predicting emerging patterns with ESPs will result in less

precise predictions compared to their prediction precisions with frequent patterns.

3. For predicting the cyclic re-occurrence of patterns from a season in its next recurring season, ESPs make high precise predictions though not significantly higher than that made by frequent patterns. As such frequent patterns can make nearly as good seasonal pattern re-occurrence predictions as ESPs.

4. For predicting the cyclic emergence of patterns, ESPs make low precise predictions, much lower than the precisions of randomly guessing the cyclic (seasonal) emergence of patterns with frequent patterns from previously occurred seasons. Hence predicting the cyclic emergence of patterns with ESPs will result in less precise predictions compared to their prediction precisions with frequent patterns.

Table 1. Notations

Notation	Meaning and/or Formula
$S^{t(i)}$	Set containing sequential patterns mined from $D^{t(i)}$
$|S^{t(i)}|$	Number of sequential patterns in $S^{t(i)}$
$E^{t(i)}_{t(j)}$	Set containing ESPs from $D^{t(j)}$ to $D^{t(i)} \mid i > j$
$|E^{t(i)}_{t(j)}|$	Number of ESPs in the set $E^{t(i)}_{t(j)}$
Pr_1	Precision of predicting $S^{t(i+2)}$ with $E^{t(i+1)}_{t(i)}$, $Pr_1 = \frac{|E^{t(i+1)}_{t(i)} \cap S^{t(i+2)}|}{|E^{t(i+1)}_{t(i)}|} \times 100\%$
Pr_2	Precision of predicting $S^{t(i+2)}$ with $S^{t(i+1)}$, $Pr_2 = \frac{|S^{t(i+1)} \cap S^{t(i+2)}|}{|S^{t(i+1)}|} \times 100\%$
Pr_3	Precision of predicting $E^{t(i+2)}_{t(i+1)}$ with $E^{t(i+1)}_{t(i)}$, $Pr_3 = \frac{|E^{t(i+1)}_{t(i)} \cap E^{t(i+2)}_{t(i+1)}|}{|E^{t(i+1)}_{t(i)}|} \times 100\%$
Pr_4	Precision of predicting $E^{t(i+2)}_{t(i+1)}$ with $S^{t(i+1)}$, $Pr_4 = \frac{|S^{t(i+1)} \cap E^{t(i+2)}_{t(i+1)}|}{|S^{t(i+1)}|} \times 100\%$
Δ_1	$Pr_1 - Pr_2$
Δ_2	$Pr_3 - Pr_4$
$D^{x,i}$	$x \in \{w=\text{winter, sp=spring, sm=summer, a=autumn}\}$ set of seasons $i \in \{1, 2\}$, seasons 1st and 2nd occurrence, D is a dataset
$Pr_{1,1}$	Precision of predicting $S^{x,2}$ with $E^{y,1}_{x,1}$, $Pr_{1,1} = \frac{|E^{y,1}_{x,1} \cap S^{x,2}|}{|E^{y,1}_{x,1}|} \times 100\%$
$Pr_{1,2}$	Precision of predicting $S^{y,2}$ with $E^{y,1}_{x,1}$, $Pr_{1,2} = \frac{|E^{y,1}_{x,1} \cap S^{y,2}|}{|E^{y,1}_{x,1}|} \times 100\%$
$Pr_{2,2}$	Precision of predicting $S^{x,2}$ with $S^{x,1}$, $Pr_{2,2} = \frac{|S^{x,1} \cap S^{x,2}|}{|S^{x,1}|} \times 100\%$
$\Delta_{1,2}$	$Pr_{1,2} - Pr_{2,2}$
$Pre_{,1}$	Precision of predicting $E^{y,2}_{x,2}$ with $E^{y,1}_{x,1}$, $Pre_{,1} = \frac{|E^{y,1}_{x,1} \cap E^{y,2}_{x,2}|}{|E^{y,1}_{x,1}|} \times 100\%$
$Pre_{,2}$	Precision of predicting $E^{y,2}_{x,2}$ with $S^{y,1}$, $Pre_{,2} = \frac{|S^{y,1} \cap E^{y,2}_{x,2}|}{|S^{y,1}|} \times 100\%$
Δ_e	$Pre_{,1} - Pre_{,2}$

Table 2. Patterns' Trend Re-occurrence Prediction

Periods	Pattern	MinSup	Patterns mined			ESPs		Pattern Prediction			ESPs Prediction		
Sales Fact Dataset													
			$|S^{t(i)}|$	$|S^{t(i+1)}|$	$|S^{t(i+2)}|$	$|E^{t(i+1)}_{t(i)}|$	$|E^{t(i+2)}_{t(i+1)}|$	Pr_1	Pr_2	Δ_1	Pr_3	Pr_4	Δ_2
		0.1	16	15	15	6	4	**100**	93.3	6.7	0.0	**26.7**	-26.7
	Closed	0.04	55	55	53	29	22	**96.6**	92.7	3.9	34.5	**40**	-5.5
$i = 1$		0.008	390	380	377	186	157	**90.9**	86.8	4.1	33.3	**41.3**	-8.0
t_1, t_2	Closed and	0.1	13	13	13	5	3	**100**	92.3	7.7	0.0	**23.1**	-23.1
	Maximal	0.04	37	40	40	23	15	**87.0**	85.0	2.0	**39.1**	37.5	1.6
		0.008	248	248	239	98	90	**83.7**	73.4	10.3	31.6	**36.3**	-4.7
		0.1	15	15	15	4	8	100	100	0.0	50.0	**53.3**	-3.3
	Closed	0.04	55	53	56	22	24	95.5	**96.2**	-0.7	27.3	**45.3**	-18.0
$i = 2$		0.008	380	377	379	157	178	**93.6**	87.5	6.1	29.3	**47.2**	-17.7
t_2, t_3	Closed and	0.1	13	13	13	3	8	100	100	0.0	**66.7**	61.5	5.2
	Maximal	0.04	40	40	42	15	17	**93.3**	92.5	0.8	26.7	**42.5**	-15.8
		0.008	248	239	256	90	101	**86.7**	78.7	8.0	26.7	**42.3**	-15.6
		0.1	15	15	18	8	15	100	100	0.0	100	100	0.0
	Closed	0.04	53	56	72	24	56	100	100	0.0	100	100	0.0
$i = 3$		0.008	377	379	565	178	362	97.5	**97.6**	-0.1	95.5	95.5	0.0
t_3, t_4	Closed and	0.1	13	13	15	8	12	87.5	**92.3**	-4.8	87.5	**92.3**	-4.8
	Maximal	0.04	40	42	49	17	33	**82.4**	78.6	3.8	82.4	**92.3**	-9.9
		0.008	239	256	351	101	163	62.3	**68.4**	-6.1	56.4	**68.4**	-12.0
		0.1	15	18	21	15	14	93.3	**94.4**	-1.1	73.3	**77.8**	-4.5
	Closed	0.04	56	72	78	56	62	**100**	98.6	1.4	**87.5**	86.1	1.4
$i = 4$		0.008	379	565	638	362	399	**98.6**	93.1	5.5	bf 79.3	70.6	8.7
t_4, t_5	Closed and	0.1	13	15	16	12	10	83.3	**86.7**	-3.4	58.3	**86.7**	-28.4
	Maximal	0.04	42	49	51	33	39	87.9	**89.8**	-1.9	78.8	**89.8**	-11.0
		0.008	256	351	382	163	182	73.0	**77.8**	-4.8	47.9	**77.8**	-29.9
		0.1	18	21	19	14	6	**100**	90.5	9.5	21.4	**28.6**	-7.2
	Closed	0.04	72	78	77	62	20	**100**	97.4	2.6	21.0	**25.6**	-4.6
$i = 5$		0.008	565	638	606	399	199	**95.7**	91.8	3.9	25.8	**31.2**	-5.4
t_5, t_6	Closed and	0.1	15	16	15	10	3	**100**	87.5	12.5	**20.0**	18.8	1.2
	Maximal	0.04	49	51	49	39	14	**100**	96.1	3.9	23.1	**27.5**	-4.4
		0.008	351	382	380	182	113	**87.9**	76.4	11.5	22.0	**29.6**	-7.6
		0.1	21	19	20	6	8	100	100	0.0	**66.7**	42.1	24.6
	Closed	0.04	78	77	74	20	36	**95**	94.8	0.2	30.0	**46.8**	-16.8
$i = 6$		0.008	638	606	607	199	290	**96.5**	90	6.5	30.7	**47.8**	-17.1
t_6, t_7	Closed and	0.1	16	15	16	3	7	100	100	0.0	**66.7**	46.7	20.0
	Maximal	0.04	51	49	47	14	24	92.9	**93.9**	-1.0	21.4	**49.0**	-27.6
		0.008	382	380	387	113	174	**89.4**	80.8	8.6	23.0	**45.8**	-22.8
								Av Δ_1		+2.66	Av Δ_2		-8.05

Table 3. Patterns' Trend Re-occurrence Prediction

Periods	Pattern	MinSup	$|S^{t(i)}|$	$|S^{t(i+1)}|$	$|S^{t(i+2)}|$	$|E^{t(i+1)}_{t(i)}|$	$|E^{t(i+2)}_{t(i+1)}|$	Pr_1	Pr_2	Δ_1	Pr_3	Pr_4	Δ_2
						TaFeng Retail Dataset							
				Patterns mined		ESPs		Pattern Prediction			ESPs Prediction		
	Closed	0.1	2	1	0	1	0	0.0	NA	0.0	0.0	NA	0.0
		0.04	55	54	53	19	31	**94.7**	92.6	2.1	52.6	**57.4**	-4.8
$i=1$		0.008	187	167	201	63	100	**98.4**	92.2	6.2	57.1	**59.9**	-2.8
t_1,t_2	Closed and	0.1	2	1	0	0	0	0.0	NA	0.0	NA	0.0	0.0
	Maximal	0.04	37	36	38	15	18	**93.3**	88.9	4.4	46.7	**50.0**	-3.3
		0.008	136	127	143	43	60	**90.7**	87.4	3.3	46.5	**47.2**	-0.7
	Closed	0.1	1	0	1	1	0	0.0	NA	0.0	0.0	NA	0.0
		0.04	54	53	50	31	6	**93.5**	88.7	4.8	**12.9**	11.3	1.6
$i=2$		0.008	167	201	163	100	42	**91.0**	81.1	9.9	20.0	**20.9**	-0.9
t_2,t_3	Closed and	0.1	1	0	1	0	0	0.0	NA	0.0	0.0	NA	0.0
	Maximal	0.04	36	38	32	17	9	**93.8**	84.2	9.6	23.5	**23.7**	-0.2
		0.008	127	143	123	78	37	**80.8**	79.7	1.1	19.2	**25.9**	-6.7
									Av Δ_1	**+3.45**	Av Δ_2		-1.48

Table 4. Patterns' Cyclic Re-occurrence Prediction

Periods	Pattern	MinSup	$|S^{w,1}|$	$|S^{sp,1}|$	$|S^{w,2}|$	$|S^{sp,2}|$	$|E^{sp,1}_{w,1}|$	$|E^{sp,2}_{w,2}|$	$Pr_{1,1}$	$Pr_{1,2}$	$Pr_{2,2}$	$\Delta_{1,2}$	$Pr_{e,1}$	$Pr_{e,2}$	Δ_e
				Patterns mined			ESPs		Pattern Prediction				ESPs Prediction		
$D^{w,1}$	Closed	0.1	16	15	18	21	6	14	100	100	100	0.0	66.7	**82.1**	-15.4
$D^{sp,1}$		0.04	55	55	72	78	29	62	100	100	100	0.0	82.8	**89.0**	-6.2
		0.008	390	380	379	565	186	399	100	100	100	0.0	74.7	**78.6**	-3.9
$D^{w,2}$		0.1	13	13	15	16	5	10	100	100	100	0.0	60.0	**77.0**	-17.0
$D^{sp,2}$	Closed and	0.04	37	40	49	51	23	39	100	100	100	0.0	87.0	**90.0**	-3.0
	Maximal	0.008	248	248	351	382	98	182	100	100	100	0.0	**86.7**	73.4	13.3
			$|S^{sp,1}|$	$|S^{sm,1}|$	$|S^{sp,2}|$	$|S^{sm,2}|$	$|E^{sm,1}_{sp,1}|$	$|E^{sm,2}_{sp,2}|$							
$D^{sm,1}$	Closed	0.1	15	15	18	21	4	6	100	100	100	0.0	**50.0**	40.0	10.0
$D^{sp,1}$		0.04	55	53	72	78	22	20	100	100	100	0.0	36.4	**37.8**	-1.4
		0.008	380	377	565	638	157	199	100	100	100	0.0	39.5	**52.8**	-13.3
$D^{sm,2}$		0.1	13	13	16	15	3	3	100	100	100	0.0	**33.3**	23.1	10.2
$D^{sp,2}$	Closed and	0.04	40	40	51	49	15	14	100	100	100	0.0	**46.7**	35.0	11.7
	Maximal	0.008	248	239	382	380	101	113	100	100	100	0.0	44.5	**47.3**	-2.8
			$|S^{sm,1}|$	$|S^{a,1}|$	$|S^{sm,2}|$	$|S^{a,2}|$	$|E^{a,1}_{sm,1}|$	$|E^{a,2}_{sm,2}|$							
$D^{sm,1}$	Closed	0.1	15	15	19	20	8	8	100	100	100	0.0	50.0	**53.3**	-3.3
$D^{a,1}$		0.04	53	56	77	74	24	36	100	100	100	0.0	41.7	**64.3**	-22.6
		0.008	377	379	606	607	178	290	100	98.9	96.0	2.9	50.0	**76.5**	-26.5
$D^{sm,2}$		0.1	13	13	15	16	8	7	100	100	100	0.0	50.0	**53.8**	-3.8
$D^{a,2}$	Closed and	0.04	40	42	49	47	17	24	100	100	100	0.0	52.9	**57.1**	-4.2
	Maximal	0.008	239	256	380	387	101	174	100	98.8	96.1	2.7	61.4	**68.0**	6.6
										Av $\Delta_{1,2}$	**+0.31**		Av Δ_e		-3.97

5 Conclusion and Future Works

ESPs detected based on current definition capture emerging trends in time-stamped datasets for decision making. Our experimental results show that ESPs improve patterns re-occurrence prediction than frequent patterns but the improvements are marginal. However, in predicting the continuous emergence of patterns, both ESPs and frequent patterns are not good predictors though the prediction precisions obtained with ESPs are slightly lower than those made with frequent patterns. The possible reasons ESPs are not good predictors for predicting the continuous emergence could arise from the fact that most ESPs detected based on current definition:

- Are not truly emerging as their emergence could be caused by noise or data fluctuations.
- Are not significantly or statistically emerging and hence not relevant in predicting the continuous emergence of patterns with time. We believe that ESPs that are significantly emerging are unlikely to have their emergence caused by noise or data fluctuations and hence could be good predictors for emerging patterns.

To help improve ESPs as good predictors of emerging patterns with time, we suggest future works on improving the existing definition of ESPs to enable detect non-trivial and more useful ESPs in time-stamped datasets. Our suggestions on future works are:

1. Incorporating statistical and/or technical methods in existing ESP definition that will employ inherent pattern properties in:
   - Validating the true emergence of sequential patterns from one dataset to another.
   - Pruning statistically insignificant emerging sequential patterns.
2. Current ESPs definition should be extended to enable detect and report concise representations of emerging sequential patterns that are interesting, non-redundant, and/or self-sufficient.

Acknowledgement. This work was partially supported by Australian Research Council Discovery grant DP130104090.

References

1. Adedoyin-Olowe, M., Gaber, M.M., Stahl, F.: TRCM: A Methodology for Temporal Analysis of Evolving Concepts in Twitter. In: Rutkowski, L., Korytkowski, M., Scherer, R., Tadeusiewicz, R., Zadeh, L.A., Zurada, J.M. (eds.) ICAISC 2013, Part II. LNCS (LNAI), vol. 7895, pp. 135–145. Springer, Heidelberg (2013)
2. Agrawal, R., Srikant, R.: Mining Sequential Patterns. In: 11th IEEE International Conference on Data Engineering, pp. 3–14. IEEE (1995)

3. Ayres, J., Flannick, J., Gehrke, J., Yiu, T.: Sequential Pattern Mining using a Bitmap Representation. In: 8th ACM SIGKDD International Conference on Knowledge Discovery, pp. 429–435. ACM (2002)
4. Boettcher, M.: Contrast and Change Mining. Wiley Interdisciplinary Reviews: Data Min. Knowl. Disc. 1(3), 215–230 (2011)
5. Chen, M.C., Chiu, A.L., Chang, H.H.: Mining Changes in Customer Behavior in Retail Marketing. Expert Syst. Appl. 28(4), 773–781 (2005)
6. Cho, Y.B., Cho, Y.H., Kim, S.H.: Mining Changes in Customer Buying Behavior for Collaborative Recommendations. Expert Syst. Appl. 28(2), 359–369 (2005)
7. Dong, G., Li, J.: Efficient Mining of Emerging Patterns: Discovering Trends and Differences. In: 5th ACM SIGKDD International Conference on Knowledge Discovery, pp. 43–52. ACM (1999)
8. Garofalakis, M.N., Rastogi, R., Shim, K.: SPIRIT: Sequential Pattern Mining with Regular Expression Constraints. In: 25th International Conference on Very Large Data Bases, pp. 7–10. Morgan Kaufmann (1999)
9. Gomariz, A., Campos, M., Marin, R., Goethals, B.: ClaSP: An Efficient Algorithm for Mining Frequent Closed Sequences. In: Pei, J., Tseng, V.S., Cao, L., Motoda, H., Xu, G. (eds.) PAKDD 2013, Part I. LNCS (LNAI), vol. 7818, pp. 50–61. Springer, Heidelberg (2013)
10. Huang, T.C.K.: Mining the Change of Customer Behavior in Fuzzy Time-Interval Sequential Patterns. Appl. Soft Comput. 12(3), 1068–1086 (2012)
11. Huang, Z., Gan, C., Lu, X., Huan, H.: Mining the Changes of Medical Behaviors for Clinical Pathways. In: 14th World Congress on Medical and Health Informatics, pp. 117–121 (2013)
12. Li, J., Liu, H., Downing, J.R., Yeoh, A.E.J., Wong, L.: Simple Rules Underlying Gene Expression Profiles of More than Six Subtypes of Acute Lymphoblastic Leukemia (ALL) Patients. Bioinformatics 19(1), 71–78 (2003)
13. Li, J., Dong, G., Ramamohanarao, K., Wong, L.: Deeps: A New Instance-Based Lazy Discovery and Classification System. Machine Learning 54(2), 99–124 (2004)
14. Li, J., Dong, G., Ramamohanarao, K.: Making Use of the Most Expressive Jumping Emerging Patterns for Classification. Knowl. Inf. Syst. 3(2), 131–145 (2001)
15. Li, J., Wong, L.: Emerging Patterns and Gene Expression Data. Genome Informatics, 3–13 (2001)
16. Liu, B., Hsu, W., Han, H.S., Xia, Y.: Mining Changes for Real-Life Applications. In: Kambayashi, Y., Mohania, M., Tjoa, A.M. (eds.) DaWaK 2000. LNCS, vol. 1874, pp. 337–346. Springer, Heidelberg (2000)
17. Mooney, C.H., Roddick, J.F.: Sequential Pattern Mining–Approaches and Algorithms. ACM Comput. Surv. 45(2), 19:1–19:39 (2013)
18. Shih, M.J., Liu, D.R., Hsu, M.L.: Mining Changes in Patent Trends for Competitive Intelligence. In: Washio, T., Suzuki, E., Ting, K.M., Inokuchi, A. (eds.) PAKDD 2008. LNCS (LNAI), vol. 5012, pp. 999–1005. Springer, Heidelberg (2008)
19. Shih, M.J., Liu, D.R., Hsu, M.L.: Discovering Competitive Intelligence by Mining Changes in Patent Trends. Trends. Expert Syst. Appl. 37(4), 2882–2890 (2010)
20. Song, H.S., Kim, S.H.: Mining the Change of Customer Behavior in an Internet Shopping Mall. Expert Syst. Appl. 21(3), 157–168 (2001)
21. Szathmary, L.: Méthodes Symboliques de Fouille de Données avec la Plate-forme Coron (Doctoral dissertation. Université Henri Poincaré-Nancy I) (2006)
22. Tsai, C.Y., Lo, C.C., Lin, C.W.: A Time-Interval Sequential Pattern Change Detection Method. International. J. Inf. Tech. Decis. 10(01), 83–108 (2011)
23. Tsai, C.Y., Shieh, Y.C.: A Change Detection Method for Sequential Patterns. Decis. Support Syst. 46(2), 501–511 (2009)

Efficient Subgraph Matching Using GPUs

Xiaojie Lin[1], Rui Zhang[1], Zeyi Wen[1], Hongzhi Wang[2], and Jianzhong Qi[1]

[1] University of Melbourne, Victoria, Australia
xiaojiel1@student.unimelb.edu.au,
{rui.zhang,zeyi.wen,jianzhong.qi}@unimelb.edu.au
[2] Harbin Institute of Technology, Harbin, China
wangzh@hit.edu.cn

Abstract. The explosive growth of various social networks such as Facebook, Twitter, and Instagram has brought in new needs for efficient graph algorithms. As a basic graph operation, subgraph matching is the foundation of many of these algorithms. Consequently, the efficiency of subgraph matching is very important and determines the speed of the whole data mining process. The development of multi-core CPUs allows subgraph matching algorithms to process multiple data at a time. However, the number of threads is still limited, which has become a bottleneck of these CPU-based algorithms. A workaround is using clusters of powerful servers, which normally incurs very expensive network transfer overhead. Therefore, improving the efficiency and parallel abilities of a single computer is a better idea. One of the most effective way to achieve this is making use of GPUs. With the ability of executing thousands of threads simultaneously, GPUs have a great potential to accelerate the subgraph matching. In this paper, we leverage the power of GPUs and propose an efficient subgraph matching algorithm. The experimental results show that our algorithm outperforms the state-of-the-art algorithm by an order of magnitude.

Keywords: Subgraph matching, GPU, relation join.

1 Introduction

Data in many different domains including social network, web, chemistry, bioinformatics etc. can be naturally modeled as graphs. These data graphs are usually complicated and large. We need fast enough data mining methods to extract useful information from them. Subgraph matching (subgraph isomorphism), which is usually a time-consuming process, plays vital roles in many of these methods. Improving the efficiency of subgraph matching is crucial in many applications. Examples include using subgraph matching to 1) locate suspicious codes in the call graphs of a program [7], 2) find protein structures that contain α-β-barrel motif [2], [10], 3) identify a small subset of molecules for further analysis in drug design [16] and 4) help social science researchers discover the relations between a successful CEO and his/her friends [18].

Much work has been done to improve the efficiency of subgraph matching, For example, the state-of-the-art STwig algorithm [14] achieves higher efficiency by

H. Wang and M.A. Sharaf (Eds.): ADC 2014, LNCS 8506, pp. 74–85, 2014.

abandoning graph structure index and combining exploration and join mechanism. However, as well as other algorithms, they all subject to the limited parallel processing ability of CPUs. Although multi-core CPUs have been developed for a long time, the number of concurrent threads is still limited, which is normally up to 16 or 20. The limited parallel ability has become a bottleneck for these CPU-based algorithms. By contrast, a high-end GPU has the ability of executing several thousand or more threads simultaneously [13], which makes it suitable for many applications involving processing a large amount of data. Compared with CPUs, GPUs also have a very high memory bandwidth. To further improve the efficiency of subgraph matching, making use of GPU is a good solution.

Our GPU-based algorithm can execute thousands of threads simultaneously to process different pieces of data. We use sophisticated memory layout in global memory to take advantage of caches and coalesced memory access. We also make full use of shared memory and constant memory to achieve extra acceleration.

1.1 Contributions and Organization of the Paper

To summarise, we make the following contributions in this paper.

- We analyse the state-of-the-art algorithm and identify bottleneck.
- We propose an efficient GPU-based subgraph matching algorithm which makes use of elaborate join order and fully pipeline mechanism.
- We conduct extensive experiments to study the performance.

The rest of the paper is organized as follows. Section 2 reviews the related work. Section 3 analyses the STwig algorithm and also briefly explains the CUDA structure. Section 4 presents our GPU-based algorithm. Section 5 presents the experimental results. Finally, Section 6 concludes this paper.

2   Related Work

2.1   Subgraph Matching

The simplest way to solve the problem of subgraph matching is brute-force search, but the computation time is unacceptable even if the graph is small. Ullmann and Julian proposed a backtracking algorithm [15] which can reduce the size of the search space significantly. In the same spirit, Cordella et al. developed a pruning based algorithm [6]. This algorithm makes use of state space representation (SSR) of the matching process and a set of feasibility pruning rules. Also, a new way of organizing data is adopted to reduce the memory requirements. This algorithm has a better time and spatial performance. However, it is still too slow and can only handle graphs with several thousand nodes.

To deal with larger graphs, a lot of algorithms use indices to achieve better performance. Systems like RDF-3X [12] and BitMat [1] create indices on distinct edges to improve performance. SpiderMine [17] mines and indexes the top-K

largest frequent patterns from the graphs. R-Join [4] uses 2-hop reachability labels [5] as its indices, and Distance-Join [18] uses a similar reachability index.

The problem of the algorithms using complex indices is obvious: the construction time or memory space for the indices are usually prohibitive. This disadvantage becomes much more significant when the graphs are very large.

The STwig algorithm [14] solves this problem by totally abandoning graph structure indices. The algorithm uses only a very simple index for mapping text labels to graph nodes, which has linear size and linear construction time. To compensate for the lack of graph structure indices, the STwig algorithm decomposes the query graph into two-level trees by a sophisticated algorithm and adopts an exploration mechanism, which can reduce both the number of two-way join operations and the size of joins' parameters. The details of STwig algorithm are discussed in Section 3.2.

To handle very large graphs, the STwig algorithm makes use of powerful clusters. The efficiency of STwig algorithm on such platforms is satisfactory. However, improving the efficiency on a single computer is still necessary. On the one hand, it is always preferable to use a single computer to get work done in time. On the other hand, less computers in a cluster are needed if a single computer can process data faster.

2.2 Join Algorithms on GPU

He et al. proposed a series of GPU join algorithms [9] including NLJ, INLJ, NINLJ and HJ. The speedup ratios for these algorithm range from 1.9X to 7.0X.

Kaldewey et al. proposed an algorithm [11] making use of zero copy mechanism, which allow the join operation to be processed "on the fly" as the GPU reads the input tables from CPU memory directly at PCI-E speed.

However, the algorithms mentioned above do not address the problem of large intermediate data produced in a multi-way join.

3   Preliminaries

3.1   Subgraph Matching

In this paper, we consider subgraph matching on a labeled graph. Let $G = (V, E, T)$ be a graph, where V is the vertex set, E is the edge set, and $T : V \rightarrow \Sigma^*$ is a labeling function assigning a label to each vertex of G. Giving a labeled query graph $q = (V_q, E_q, T_q)$, subgraph matching will find all occurrences of q in G.

Take the query graph and data graph shown in Fig 1(a)(c) as an example, the matching results are $(A_1, B_1, C_1, D_1, E_1)$.

3.2   The STwig Algorithm

STwig algorithm [14] can be divided into steps as described below.

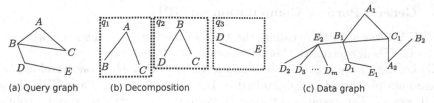

Fig. 1. A data graph and a query graph

Query Graph Decomposition is the first step. a query graph will be decomposed into a set of basic query units called STwigs. A STwig is a two-level tree, whose structure is very simple so that matching it in the data graph is very easy.

An example of the decomposition is shown in Fig 1(a)(b). The decomposition algorithm tries to minimize the number of STwigs and ensure that we can find a STwig order $\langle o_1, o_2, \cdots, o_n \rangle$ which satisfies the following statement: $\forall i >= 2$, $\exists j < i$ such that the root node of STwig o_i is the same node as a leaf of STwig o_j. $\langle q_1, q_2, q_3 \rangle$ is such an order in the example of Fig 1. The order can be used in the step of matching STwigs.

Matching STwigs in the data graph is simple. The system constructs a label index at the beginning so that nodes with a specified label can be retrieved directly. The algorithm of matching the first STwig o_1 is shown in Algorithm 1.

Algorithm 1. Match the First STwig

---

input : Data graph G, STwig $o_1 = (r_1, L_1)$ where r_1 is the root node and L_1 is the leave label set

output: Result set R

Fill the set S_1 with all nodes with r_1's label;

for *each n in S_1* **do**

    for *each l_i in L_1* **do**

        $T_{l_i} \leftarrow \{m | m \in n.\text{neighbors and } m.\text{label}=l_i\}$;

    $R = R \cup \{\{n\} \times T_{l_1} \times T_{l_2} \times \cdots \times T_{l_{|L_1|}}\}$;

The matching algorithm for other STwigs is slightly different because some pruning mechanism can be used. Let us consider Fig 1. The matching order is $\langle q_1, q_2, q_3 \rangle$. The result sets for q_1 and q_2 are $R_1 = \{(A_1, B_1, C_1), (A_2, B_2, C_1)\}$ and $R_2 = \{(B_1, C_1, D_1)\}$ respectively. When q_3 is being matched, the candidate set for its root node is not $\{D_1, D_2, \cdots, D_m\}$. R_2 ensures that only node D_1 in the data graph is possible to match node D in the query graph, so the candidate set for q_3 is $\{D_1\}$ and $R_3 = \{(D_1, E_1)\}$. This pruning strategy can reduce memory usage and the workload of the following join operation significantly.

Joining STwig Results is the final step. For the example above, the final result set is $R_1 \bowtie R_2 \bowtie R_3 = \{(A_1, B_1, C_1, D_1, E_1)\}$

3.3   General Purpose Computation on GPU

A CPU invokes a GPU by calling the *kernel* function, then the multiprocessors of the GPU will execute the kernel function simultaneously.

GPU threads are organized in a two-level hierarchy. The lower level is *block*, where multiple threads are organized in 1D, 2D or 3D ways. Each GPU thread has it own ID. For example, a thread ID comprises a row number and a column number in a 2D block. The higher level is *grid*, which contains multiple blocks.

GPUs also have a memory hierarchy. A single thread has a private memory space provided by registers and *local memory*. Registers is very fast but the amount is limited and they cannot be used to store an array indexed with non-constant quantities. Local memory resides in big but slow *device memory* and it does not have such restriction. Threads in the same block share *shared memory*, which is on-chip, small and very fast. All threads on a GPU can access to *global memory*, *constant memory* and *texture memory*, all of which reside in device memory. Constant memory and texture memory have dedicated caches to speed up the access. GPUs with *compute capability* higher than 2.0 have L1 and L2 caches for global memory, which is similar to CPUs.

4   Our Proposed GPU-Based Algorithm

Not all of the 3 steps of the original STwig algorithm are perform on the GPU.

The decomposition step consumes only a very small amount of time, so there is no need to perform it on the GPU. The step of matching STwig is also performed on the CPU because: 1) The computation time is not as long as that of the join step normally; 2) The data graph may occupy too many GPU memory; 3) We can pipeline the CPU steps and the GPU steps to achieve extra speedup.

Joining STwig results is the step on which we focus. In the worst case, each single 2-way join can produce a result set whose size is the product of input relation sizes, so both the computation time and memory usage are problems.

We propose a GPU-based join algorithm in the following subsections to solve these problems. The algorithm makes use of hash table to improve the efficiency. Although the random memory access pattern of hash table will negatively affect GPU memory bandwidth, the constant search time still makes it a good choice.

4.1   Choosing a Join Tree

We denote the result sets of all STwigs by $R_1, R_2, \cdots, R_n$. A multi-way join can be represented by a join tree. Each leaf of the join tree represents the result set of a STwig. Each internal node represents a 2-way join. The left child of an internal node is the *build relation*, which means a search data structure will be built for it. The right child is the *probe relation*, whose tuples are used to matche with those of the build relation.

Two different join trees are shown in Fig 2 ($\langle p_1, p_2, \cdots, p_n \rangle$ is a permutation of $\{1, 2, \cdots, n\}$). GPU memory is limited and cannot be enlarged freely, so the major consideration of choosing join tree is memory usage.

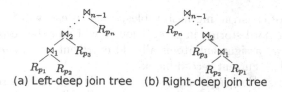

(a) Left-deep join tree   (b) Right-deep join tree

Fig. 2. Join tree

Let us consider the left-deep join tree. To save GPU memory, we only load R_{p_1} and build the corresponding hash table H_1 in the GPU at first. Then GPU threads read tuples of R_{p_2} from CPU memory directly, conduct the first 2-way join and store results in the GPU. After the join, H_1 can be released and hash table H_2 will be built for the result set of $R_{p_1} \bowtie R_{p_2}$ in the GPU. The second 2-way join $(R_{p_1} \bowtie R_{p_2}) \bowtie R_{p_3}$ can then begin. Consequently, only one hash table and one intermediate result buffer reside in GPU. However, they may still be too large. In the worst case, $|R_{p_1} \bowtie R_{p_2}|$ may be as large as $|R_{p_1} \times R_{p_2}|$. Workarounds for the memory problem like using block-based mechanism exists. However, they can slow down the process significantly. *Bushy join trees* (trees that are neither left-deep or right-deep) have the same problem.

The right-deep join tree is a better idea because a fully pipeline mechanism can be used. At the beginning, hash tables $H_2, H_3, \cdots, H_n$ will be constructed in the GPU for $R_{p_2}, R_{p_3}, \cdots, R_{p_n}$ respectively. Because of the pruning mechanism used in matching STwigs, $R_2, R_3, \cdots, R_n$ are much smaller than R_1. In many cases, $|\sum_{i=2}^{n} R_i|$ is even smaller than $|R_n|$, so we can let R_1 to be R_{p_1} to save a large amount of memory. Besides, we use a fully pipeline mechanism to minimize memory space for the intermediate data, which is described in Section 4.3.

The Order of Leaves. Further improvement can be achieved by optimizing the order of leaves, i.e. optimizing the permutation of $\langle p_1, p_2, \cdots, p_n \rangle$. Better order can result in less intermediate results which means less join time. To choose the optimal order, we can predict the size of intermediate results [8].

4.2   Hash Tables on GPU

Definitions of Operators. The hash table operators defined here will be used to explain our join algorithm later. H_i.`match`(x) returns the iterator pointing to the first record that match x. $iterator$.`record`$()$ returns the record and $iterator$.`next`$()$ returns the iterator pointing to the next record that match x.

Implementation Details. We implement the hash table by chaining mechanism, which can be constructed in parallel easily. Data is stored together in two arrays to avoid memory overheads. The first array is *pool*. Each record of $R_{p_i}(1 < i \le n)$ occupies $l_i + 1$ successive entities of *pool*, where l_i is the size of a R_{p_i}'s record. The *extra entity* is used to store the index of next record with the same hash value. Another array is *head*. Each entity of *head* stores the index of the first record in the chain with corresponding hash value.

In the phase of constructing hash tables, each thread will first calculate the hash value of a record storing in *pool*. Then it will link the record at the head of the chain by two assignment atomically: 1) extra entity$\leftarrow$ *head*[hash value] 2) *head*[hash value] $\leftarrow$ current record index.

We use the least $\lfloor \log |R_{p_i}| \rfloor$ significant bits of the sum of a record's entities as the hash value, so the size of array *head* is $2^{\lfloor \log |R_{p_i}| \rfloor}$, and the size of array *pool* is $|R_{p_i}|(l_i + 1)$. The total size of a hash table is $2^{\lfloor \log |R_{p_i}| \rfloor} + |R_{p_i}|(l_i + 1)$. Apparently, $2^{\lfloor \log |R_{p_i}| \rfloor} \leq |R_{p_i}|$, so

$$\frac{\text{hash table size}}{\text{data size}} = \frac{2^{\lfloor \log |R_{p_i}| \rfloor} + |R_{p_i}|(l_i + 1)}{l_i |R_{p_i}|} \leq 1 + \frac{2}{l_i}$$

Because $l_i \geq 2$, a hash table occupies at most 2 times memory.

4.3 Joining STwig Results

Our pipeline joining algorithm (see Algorithm 2) works like a depth-first search and only a very small and fixed amount of intermediate data need to be stored. The explanation of the algorithm is separated in the following subsections.

Algorithm 2. Join algorithm for each thread

    Input: Probe relation R_{p_1} and hash tables $H_2, H_3, \cdots, H_n$
1 **begin**
       // *pos* indicates this thread is processing subjoin $\bowtie_{pos}$
2      $pos \leftarrow 1$;
3      $itArray[1] \leftarrow none$;
4   **while** *true* **do**
5   **while** $itArray[pos] = none$ *and* $pos \neq 1$ **do**
6             $pos \leftarrow pos - 1$;
7   **while** $itArray[1] = none$ **do**
8             $i \leftarrow$ GetNewRecordsOrQuit();
9   set *imResult* based on $R_{p_1}[i]$;
10            $itArray[1] \leftarrow H_2$.match(*imResult*);
11   set *imResult* based on *pos* and $itArray[pos]$.record();
12         $itArray[pos] \leftarrow itArray[pos]$.next();
13   **if** $pos = n - 1$ **then**
14   output *imResult* as a final result;
15   **else**
16            $itArray[pos + 1] \leftarrow H_{pos+2}$.match(imResult);
17            $pos \leftarrow pos + 1$;

Definitions of Variables and Functions. Each thread of the GPU has two arrays. One is $imResult[1, 2, \cdots, q_n]$, which stores the intermediate result. Each entity corresponds to one query graph node, so a final joining result is produced when all entities are set. Another is $itArray[1, 2, \cdots, n-1]$. When a thread is processing subjoin $\bowtie_i$ (see the join tree in Fig 2(b)), it stores an iterator of H_{i+1} in $itArray[i]$. Both $imResult$ and $itArray$ are frequently accessed and small enough, so we put them in shared memory. Each thread has a variable pos which is used to indicate the subjoin being processed. For example, the thread will be processing $\bowtie_1$ if $pos = 1$.

The function GetNewRecordsOrQuit() is used to retrieve the first unhandled record of R_{p_1}. If there is no unhandled record, this function will terminate the thread.

A DFS-Like Procedure. The strategy of our algorithm is to "consume" the intermediate results as soon as possible. In the whole process of the join, a GPU thread will move forward to process next subjoin whenever possible and only moves backward when the current subjoin fails. The manner is similar to that of a depth-first search algorithm.

An Example. We use an example to illustrate how our algorithm works.

Let us consider the data graph and query in Fig 1 again. The result sets of STwigs and the final result set are $R_1 = \{(A_1, B_1, C_1), (A_2, B_2, C_1)\}$, $R_2 = \{(B_1, C_1, D_1)\}$, $R_3 = \{(D_1, E_1)\}$ and $R_1 \bowtie R_2 \bowtie R_3 = \{(A_1, B_1, C_1, D_1, E_1)\}$ respectively.

In this example, the probe relation is R_1 and we first join R_2 with R_1, so $\bowtie_1 = R_2 \bowtie R_1$ and $\bowtie_2 = R_3 \bowtie (R_2 \bowtie R_1)$. We assume $imResult[1]$ corresponds to node A of query graph; $imResult[2]$ corresponds to node B ...

At the beginning, the GPU constructs hash tables H_2 and H_3 for R_2 and R_3 respectively. Then the join process starts. To simplify the explanation, we assume there is only one thread.

When Algorithm 2 starts, the thread will first initialize pos to 1 and $itArray[1]$ to *none*. Then it starts the first round of the outermost block of while.

In line 8, it calls GetNewRecordsOrQuit() to get the index of the first unhandled record of R_1. Then $imResult$ is set to $(A_1, B_1, C_1, -, -)$ (line 9). The match operator of H_2 checks the second and third entities of $imResult$ and returns an iterator pointing to record (B_1, C_1, D_1) of R_2 to set $itArray[1]$ (line 10). Based on the value of $itArray[1].record()$, $imResult$ is further updated to $(A_1, B_1, C_1, D_1, -)$ (line 11). $itArray[1]$ is updated again to point to next record in line 12. It is actually updated to *none* because there is no other record in R_2 whose first two entities are B_1 and C_1. Preparing for next round, $itArray[2]$ is set to an iterator pointing to the only matching record of R_3, (D_1, E_1), in line 16, and pos is increased to 2 in line 17.

In the second round, $imResult[5]$ will be set to E_1 in line 11. At this point, all entities of $imResult$ are set properly, so the thread will output the final result $(A_1, B_1, C_1, D_1, E_1)$ contained in $imResult$ in line 14.

At the third round, *pos* will be decreased to 1 again in the while block beginning in line 5. Then in the while block beginning in line 7, it calls `GetNewRecordsOrQuit()` to get a new record of R_1, i.e., (A_2, B_2, C_1), to set *imResult*. However, no record of R_2 match *imResult* this time and *itArray*[1] is set to *none* (line 10). The thread calls `GetNewRecordsOrQuit()` again and terminates itself because no record is unhandled.

4.4 Storage

Storage of R_{p_1}. The result set R_{p_1} can be stored in either CPU or GPU and both solutions require the memory access to be coalesced, so we store R_{p_1} in a well-padding 2D array. Each record of R_{p_1} will be stored in one column.

The Shared Memory. *imResult* and *itArray* cannot be stored in registers because they are not indexed with constant quantities, so we store them in shared memory as mentioned above. To avoid serializing access due to bank conflicts, 32 *imResults* of all threads in the same warp are stored together in a 2D array. Each *imResult* occupies one column so that each thread can always access *imResult* via its own bank. *itArray* is stored in the same way.

5 Experimental Study

All the experiments were conducted on a computer with 8GB RAM, a 3.30GHz Intel(R) Xeon(R) E5-2643 CPU and a Tesla C2075 GPU. Tesla C2075 has 5GB global memory and its compute capability is 2.0.

We use synthetic data in the following experiments to study how the properties of data graphs and query graphs affect our speedup ratio. The data graphs in this section are generated by using R-MAT [3] model. The graphs are undirected and the parameters of R-MAT model are $a = 0.4$, $b = c = d = 0.2$. A query graph with N nodes is generated by adding $2N$ edges randomly.

5.1 Computation Time of Each Steps

To analyze the efficiency of each step, we conducted an experiment and recorded the computation time. The data graph we used in this experiment has 16K nodes and 16 labels. The average degree is 32. We generated 100 query graphs whose numbers of nodes range from 5 to 10. The results are shown in Fig 3a.

The total time is the sum of "(Copy &) Hash" time, join time and other time consumed by common operations which are the same both CPU-based and GPU-based algorithms. We can see that the join time normally dominates the total time, so our approach of optimizing the join operation achieves an overall speedup of an order of magnitude. Actually, if we only consider the join time, the speedup ratio reaches to 26.0. On the other hand, the GPU takes more time to finish the step of "(Copy &) Hash", but this time is relatively small and will

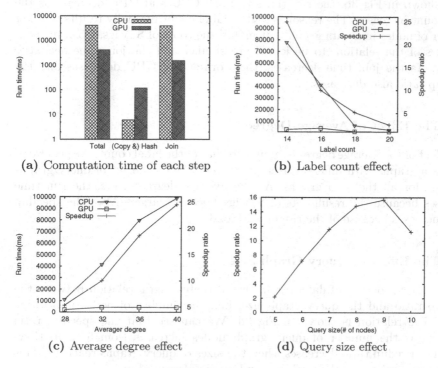

(a) Computation time of each step

(b) Label count effect

(c) Average degree effect

(d) Query size effect

Fig. 3. Experiment Results

not affect the overall speedup ratio. The major reasons for this phenomenon are 1) the GPU has to copy data from CPU memory before building hash tables, which can consume considerable time, 2) our GPU algorithm uses lots of atomic operations in the step of building hash tables and 3) the uncoalesced memory access pattern of this step can slow down GPU algorithm.

5.2 The Effect of Data Graph Size

To verify the performance of GPU algorithm with different size of data graph, we conducted some experiments in which only the numbers of nodes of data graphs are changed. The numbers of labels and the average degrees for all data graphs are 16 and 32 respectively. The numbers of data graph nodes are 16K, 32K, 64K and 128K. Our GPU algorithm outperforms CPU in all cases, and the speedup ratios range from 7.8 to 16.7.

5.3 The Effect of Label Count

In this experiment, we study the effect of the number of labels. We used several data graphs with the same node counts (16K) and average degrees (32). Only the label counts are different.

As shown in Fig 3b, the run times of both CPU and GPU decrease as the label count increases. The reason is higher label count normally results in fewer number of nodes in a data graph that match the roots of STwigs. Consequently, the sizes of the relations to be joined are smaller and the join time decreases rapidly. As the join time decreases, the workload of GPU decreases and the speedup ratio also decreases.

5.4   The Effect of Average Degree

The effect of the average degree is similar to that of the label count. We generated four data graphs with different average degrees. The same set of query graphs are used for all the data graphs. As the average degree grows, the run time increases because the result sets of STwigs becomes larger. Also, the speedup ratio increases because of the rise of workload.

5.5   The Effect of Query Graph Size

Besides the properties of data graphs, we also observed a relation between the speedup ratio and the query graph size, i.e., the number of nodes in a query graph. The relation is shown in Fig 3d. We can see that the speedup ratio increases as the number of query graph nodes increases from 5 to 9. However, the speedup ratio decreases when the sizes of query graphs reach 10. This phenomenon relates to the workload and our implementation.

When the query graph is small, smaller number of STwigs and fewer edges in each STwig results in light workload. In such cases, we cannot make full use of the powerful computation abilities of the GPU. When the query graphs are larger, the workload increases and each GPU thread can keep itself busy. Consequently, the average speedup ratio for query graphs with 9 nodes reaches 15.66. However, if a query graph is too large, the speedup ratio may decline because $imResult$ and $itArray$ occupied too much shared memory and thus the number of the concurrent threads will decline. In our experiment, concurrent thread counts can drop from 21504 to 10752.

6   Conclusions

Subgraph matching involves a large amount of data to be processed. Consequently, the computation time is very long. To alleviate this problem, parallel technology is widely used. However, most of the approaches only try to exploit CPUs' parallel abilities, which is very limited. By contrast, our algorithm makes use of GPUs. A high-end GPU has the ability of executing thousands of threads simultaneously. This ability allows our algorithm to outperform the state-of-the-art algorithm by an order of magnitude based on the experimental results. Additionally, we observe that our approach can perform even better in a heavy workload situation.

References

1. Atre, M., Chaoji, V., Zaki, M.J., Hendler, J.A.: Matrix bit loaded: A scalable light weight join query processor for rdf data. In: Proceedings of the 19th International Conference on World Wide Web, pp. 41–50. ACM (2010)
2. Branden, C., Tooze, J., et al.: Introduction to protein structure, vol. 2. Garland, New York (1991)
3. Chakrabarti, D., Zhan, Y., Faloutsos, C.: R-mat: A recursive model for graph mining. Computer Science Department, 541 (2004)
4. Cheng, J., Yu, J.X., Ding, B., Yu, P.S., Wang, H.: Fast graph pattern matching. In: IEEE 24th International Conference on Data Engineering, ICDE 2008, pp. 913–922. IEEE (2008)
5. Cohen, E., Halperin, E., Kaplan, H., Zwick, U.: Reachability and distance queries via 2-hop labels. SIAM Journal on Computing 32(5), 1338–1355 (2003)
6. Cordella, L.P., Foggia, P., Sansone, C., Vento, M.: A (sub) graph isomorphism algorithm for matching large graphs. IEEE Transactions on Pattern Analysis and Machine Intelligence 26(10), 1367–1372 (2004)
7. Eichinger, F., Böhm, K., Huber, M.: Mining edge-weighted call graphs to localise software bugs. In: Daelemans, W., Goethals, B., Morik, K. (eds.) ECML PKDD 2008, Part I. LNCS (LNAI), vol. 5211, pp. 333–348. Springer, Heidelberg (2008)
8. Garcia-Molina, H., Ullman, J.D., Widom, J.: Database system implementation, vol. 654. Prentice Hall Upper Saddle River, NJ (2000)
9. He, B., Yang, K., Fang, R., Lu, M., Govindaraju, N., Luo, Q., Sander, P.: Relational joins on graphics processors. In: Proceedings of the 2008 ACM SIGMOD International Conference on Management of Data, pp. 511–524. ACM (2008)
10. He, H., Singh, A.K.: Graphs-at-a-time: Query language and access methods for graph databases. In: Proceedings of the 2008 ACM SIGMOD International Conference on Management of Data, pp. 405–418. ACM (2008)
11. Kaldewey, T., Lohman, G., Mueller, R., Volk, P.: Gpu join processing revisited. In: Proceedings of the Eighth International Workshop on Data Management on New Hardware, pp. 55–62. ACM (2012)
12. Neumann, T., Weikum, G.: The rdf-3x engine for scalable management of rdf data. The VLDB Journal 19(1), 91–113 (2010)
13. NVIDIA: CUDA C best practices guide (2013)
14. Sun, Z., Wang, H., Shao, B., Li, J.: Efficient subgraph matching on billion node graphs. Proceedings of the VLDB . . . , 788–799 (2012)
15. Ullmann, J.R.: An algorithm for subgraph isomorphism. Journal of the ACM (JACM) 23(1), 31–42 (1976)
16. Yan, X., Yu, P.S., Han, J.: Substructure similarity search in graph databases. In: Proceedings of the 2005 ACM SIGMOD International Conference on Management of Data, pp. 766–777. ACM (2005)
17. Zhu, F., Qu, Q., Lo, D., Yan, X., Han, J., Yu, P.: Mining top-k large structural patterns in a massive network. Proceedings of the VLDB Endowment 4(11) (2011)
18. Zou, L., Chen, L., Özsu, M.T.: Distance-join: Pattern match query in a large graph database. Proceedings of the VLDB Endowment 2(1), 886–897 (2009)

A Negative-Aware and Rating-Integrated Recommendation Algorithm Based on Bipartite Network Projection

Fengjing Yin, Xiang Zhao, Guangxin Zhou, Xin Zhang, and Shengze Hu

National University of Defense Technology, Changsha, 410073, P.R.China
{yinfengjing,xiangzhao,gxzhou,ijunzhang,szhu}@nudt.edu.cn

Abstract. Bipartite network projection method has been recently employed for personal recommendation. It constructs a bipartite network between users and items. Treating as resource in the network user taste for items, it allocates the resource via links between user nodes and item nodes. However, the taste model employed by existing algorithms cannot differentiate "dislike" and "unrated" cases implied by user ratings. Moreover, the distribution of resource is solely based on node degrees, ignoring the different transfer rates of the links. To enhance the performance, this paper devises a negative-aware and rating-integrated algorithm on top of the baseline algorithm. It enriches the current user taste model to encompass "like", "dislike" and "unrated" information from users. Furthermore, in the resource distribution stage, we propose to initialize the resource allocation according to user ratings, which also determines the resource transfer rates on links afterward. Extensive experiments conducted on real data validate the effectiveness of the proposed algorithm.

1 Introduction

With the rapid growth of the World Wide Web, people are emerged in an overwhelming amount of information, which makes it difficult to obtain relevant information of interest. Personal recommendation is employed to suggest products to the consumers who may be interested in, such as news, books, music, movies, and etc. As a consequence, a large number of diverse algorithms were proposed to solve the problem. Besides the classic methods, e.g., content-based methods [1], collaborative filtering and variants [2-4], some new paradigms have been introduced lately, including matrix factorizations [5,6], social filtering [7] and network based [8-10].

Bipartite network projection was initially introduced in physics, but found applications in personal recommendation [8]. It relies on a resource distribution process in a bipartite network to provide a top-n recommendation. Particularly, it considers users and items as two types of network nodes, respectively, and treats user taste as resource to be allocated in the bipartite network. User rating is scaled into a numeral (either 1 or 0), in comparison with the median of the pre-defined rating range, to indicate user taste. In another word, numeral 1 means the user likes a particular item, while numeral 0 means the user does not like or has not rated the item. Afterwards, two rounds of resource distribution are carried out, regardless of the user ratings.

H. Wang and M.A. Sharaf (Eds.): ADC 2014, LNCS 8506, pp. 86–97, 2014.

Eventually, the final resource that an unrated item gets indicates its possibility to be recommended to users. The method was shown to outperform the collaborative filtering methods. Albeit, we observe that there are at least two shortcomings that limit its further improvement. Firstly, the current user taste model does not differentiate "dislike" and "unrated" cases, both expressed by numeral 0, which potentially stops the algorithm to provide a more precise recommendation. Secondly, the resource distribution process does not leverage user ratings; that is, the transfer rates on different links are not proportional to the corresponding user ratings. Such allocation, and hence the recommendation, can be inaccurate when a user has a biased preference among the items. We rectify these issues in this work.

This paper first presents a *negative-aware* user taste model to encompass "like" "dislike" and "unrated" cases implied by user ratings so that the user preference is fully reflected. Instead of using the pre-defined rating range, we propose to use an adaptive user rating range as threshold for determining user taste. While the rationale behind is similar to the existing model, we compare a rating with the *average* rating of the user. Hence, a rating above this average is considered as a "like", expressed by numeral 1; a rating below this average is regarded as a "dislike", expressed by numeral -1; and unrated cases are noted by numeral 0. Based on this user taste model, we initialize the resource of each item according the user taste and the proportion of the rating to the sum of all ratings from this user, which can appropriately reflect the dissimilarity of user interest. Similarly, when deriving the transfer rates of links in resource distribution, we design a *rating-integrated* method to allocate resource via a link *proportional* to the user ratings on the corresponding item. In this way, the final resource allocation is expected to more expressively reflect user preference. Further, we validate the effectiveness of the model and method on real data in comparison with four existing algorithms based on bipartite network projection.

To summarize, we make the following contributions:

- We devise a negative-aware user taste model to encompass "like" "dislike" and "unrated" cases implied by user ratings;
- We propose a rating-integrated method to allocate initial resource and determine transfer rates on network links according to user ratings;
- The proposed model and method constitute a new algorithm for personal recommendation, which is demonstrated to outperform alternatives through extensive experiments on public real data.

The rest of the paper is organized as follows: Section 2 introduces the preliminaries and the baseline algorithm. Section 3 presents the new taste model and the method for initialization and distribution of the resource. Experimental results are shown in Section 4. We discuss the related work in Section 5, followed by conclusion in Section 6.

2 Preliminaries

The baseline algorithm for personal recommendation based on bipartite network projection relies on a bipartite network, consisting of two types of nodes - *user* and *item* nodes, denoted by U and V, respectively. Let u_i denote the i-th user in U, and

v_j denote the j-th item. u_i has a rating $r(u_i, v_j)$ on v_j. All the ratings from user u_i constitute a rating set $R(u_i)$ with cardinality of $|R(u_i)|$, and all the ratings on item v_j constitute a rating set $R(v_j)$ with cardinality of $|R(v_j)|$. *Links* exist between different types of nodes. That is, a link always connects a user node and an item node. Each link models a rating behavior of a user on an item. Assume every rated item by a given user is assigned with certain quantity of resource, i.e., *initial resource* allocation. The resource first flows from item nodes to user nodes, then back to item nodes along the links. Hence, every item gets a *final resource* allocation after the *two-round resource distribution*. The top n items with bigger resource would be recommended to the current user then the recommendation will be executed for the next user.

Consider the initial resource as recommendation power. The intuition behind is that if an item gets more resource after distribution, it is more highly to be liked based on the user's preference. Thus, the resource distribution process is of importance. Currently, the taste of a user is determined based on a threshold equal the median of the pre-defined rating range. Specifically, a user likes an item, expressed by numeral 1, if she rates the item above the threshold; otherwise, the case is concluded into the unrated category, expressed by numeral 0. For example, the threshold is 3, if the predefined rating range is $[1,5]$; a user rating 4 on an item implies that she likes the item; a user rating 2 on an item is diminished to unrated case.

After determining user tastes, the resource allocation process in the bipartite network is carried out. Two rounds of resource distribution are conducted. In the first round, resources are transferred from item nodes to user nodes, and all user nodes having links to a particular item share its resource equally. In the second round, resources are transferred back to item nodes, and all item nodes having links to a user share its resource equally. The final resource allocated to each item indicates its probability of being recommended to the given user. For a top-n recommendation, a list of n unrated items with largest resource allocation is created for the given user. We refer this as "the baseline algorithm" in the rest of the paper when context is clear.

Example 1: Fig. 1 shows an example of recommendation for user u_2 by the baseline algorithm. Assume in Fig. 1(a) is the constructed bipartite network, with item nodes in the top and user nodes in the bottom. The ratings of user u_2 to items v_1 to v_5 are 5, 0, 2, 0 and 4, respectively, and the median is 3. So the initial resource configuration is 1, 0, 0, 0 and 1, respectively. The initial resource and raw rating value are tagged above the item nodes. In the first round as in Fig. 1(b), resources are transferred from item nodes to users nodes; e.g., user u_2 gets $1/3$ from v_1, and $1/3$ from v_5, $2/3$ in total. Then, in the second round as in Fig. 1(c), resources are transferred back to item nodes; e.g., item v_2 receives $1/6$ from users u_1 and $1/12$ from user u_3. So the final resource allocation for item v_2 is $1/4$ in total. Similarly, item v_4 gets $11/36$ in total. So item v_4 will be recommended to user u_2 for its allocated resource is the biggest among the unrated items of user u_2.

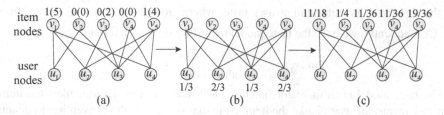

Fig. 1. Example of resource distribution in baseline algorithm

3   A Negative-Aware and Rating-Integrated Algorithm

This section improves the baseline by introducing a negative-aware and rating integrated algorithm for personal recommendation.

3.1   User Taste Model

Let us take a motivating example.

Example 2: Consider a pre-defined rating range [1, 5], and a user rating 1 on an item. In the baseline algorithm, we take the median of the rating range, i.e., 3 as the threshold. As 1 is below the threshold, the user may not like the item. Subsequently, we use a numeral 0 in the initial resource allocation for this item. Recall that a user taste to an item is also expressed by 0 if she has not rated the item.

The current user taste model does not distinguish the aforementioned cases, which hinders the algorithm from reaching a higher recommendation precision. Furthermore, it is intuitive that a rating close to the bottom of the rating range implies a low satisfaction or preference towards an item, i.e., negative attitude towards the item. We argue that the rating reflects the user's taste only if it is compared with her own rating range, rather than the pre-defined rating range. As a consequence, we distinguish the three cases -"like" "dislike" and "unrated" - in the new *adaptive* user taste model.

Firstly, we adopt as the reference the user's rating range, instead of the pre-defined rating range. That is, we compare a user rating with respect to the user's own rating range; e.g., given a pre-defined rating range [1, 5], a user always rates within [2, 4], then the latter is the user's own rating range. This is intuitive, as users may have different rating habit. Some users are harsh when rating, and hence, always give ratings across the whole pre-defined rating range; while some are soft towards the items, and hence, usually gives ratings within a small sub domain of the pre-defined rating range.

Subsequently, it is further observed that, as to an item, even if her rating is above some others', this may indicate that she dislike the item, since she has a narrow ranting range. Therefore, instead of using median of the pre-defined rating range, we propose to use the average of a user's ratings as the threshold to determine the user's taste towards an item. That is, we take

$$\bar{r}(u_i) = \frac{\sum\limits_{v_j \in V} r(u_i, v_j)}{|R(u_i)|}$$

as the threshold for determining user taste, where $R(u_i)$ is the set of ratings from u_i. Consequently, we have the taste model regarding user u_i and item v_j:

$$T(u_i, v_j) = \begin{cases} 1, & r(u_i, v_j) \geq \bar{r}(u_i) \\ -1, & otherwise \end{cases}.$$

(1)

The taste model is an indicator function such that 1 means the user likes the item, while -1 means the user dislike the item. Note that we follow the convention to denote as 0 the user's taste for an unrated item.

Example 3: Consider a pre-defined rating range of [1, 5], and a user rates five items as 2, 3, 3, 4, and 5, respectively. Thus, according to the new model, we first take the average $(2+3+3+4+5)/5 = 3.4$ as the threshold. Hence, rating 3 is considered as a dislike, since it is close to the bottom of the rating range according the user's rating habit. On the other hand, rating 4 implies a like, since it is above the threshold.

Our user taste model is adaptive in the sense that it adjusts the threshold according to different users so that the user's taste can be well reflected. Introducing the taste of dislike can make full use of user taste information. Moreover, this distinguishes the dislike and unrated cases that are treated identically in the baseline, and hence the recommendation performance is expected to be improved due to these improvements. We will see shortly how this affects the resource distribution process.

3.2   Initial Resource Allocation

A higher rating implies a stronger recommendation from a user towards an item. To reflect this information in the initial resource allocation, we weight the initial resource for every item with a coefficient. In particular, we multiple the taste numeral with the ratio of the user's rating to current item to the average of the user's ratings for all items. For a given user u_i and an item v_j, the initial resource allocated to v_j is

$$R_{ini}(v_j) = \hat{r}(u_i, v_j) \times T(u_i, v_j),$$

(2)

where $T(u_i, v_j)$ is as the taste model formulated in Section 3.1, $\hat{r}(u_i, v_j)$ denotes the weight we put on the user taste to generate the initial resource, which is

$$\hat{r}(u_i, v_j) = \begin{cases} \dfrac{r(u_i, v_j)}{\bar{r}(u_i)} & if\,(r(u_i, v_j) > \bar{r}(u_i)) \\ \dfrac{tr - r(u_i, v_j)}{\bar{r}(u_i)} & otherwise \end{cases},$$

(3)

where tr is the top of the user rating range, $tr - r(u_i, v_j)$ is to make sure that a smaller rating will get a smaller initial resource under the negative taste when the rating is less than the average of user ratings.

This initial resource allocation tries to emphasize the distinction of user taste to different items that the user likes. Thus, the initial resources become more distinguishable and accordant with user taste.

Example 4: In the baseline algorithm, the initial resource is the same as the user taste for items. Since the ratings of the user to the items with same user taste are different, we have good reasons to doubt the way of initializing resource by simply equating the initial resource with the user taste for them. In Fig. 2(a), the numbers next to the links are the ratings. To use the aforementioned model to allocate initial resource, we first determine the user taste for u_2 by formula (1), as 1, 0, -1, 0 and 1, respectively, for all items. Then, we allocate initial resource by weighting 5/3.7, 2/3.7, and 4/3.7 for the non-zero user taste respectively, as formulae (2) and (3) represented.

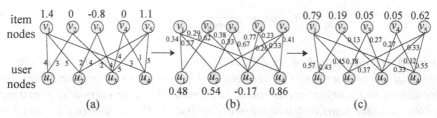

Fig. 2. Example of resource distribution in proposed algorithm

3.3   Resource Distribution

Resource distribution based degree of nodes ignores the difference of links and cannot reflect the distinctness in the extent of user interest to the same item. We make resource transfer from item nodes to user nodes and back to items both aware of the ratio of user rating to the sum of ratings from all users who have rated the given item or the sum of ratings by the user to all items. This can improve the accuracy of resource distribution to make the distribution process more specific. In particular, in the first round of transfer from item nodes to user nodes, the ratings are normalized with each user's average rating to avoid the influence of rating bias from different users. Thus, for each user node u_i, the resource transferred from item v_j to this node is

$$R(u_i, v_j) = \frac{\hat{r}(u_i, v_j)}{\sum\limits_{u_k \in U} \hat{r}(u_k, v_t)} \times R_{\text{ini}}(v_j). \tag{4}$$

Adding the contributions from all item nodes connecting to u_i,

$$R(u_i) = \sum_{v_j \in V} R(u_i, v_j).$$

Example 5: Recall Example 4. In Fig. 2(b), the numbers around the edges are the transfer rates from item nodes to user nodes. The initial resource is distributed from item nodes to user nodes in the first round by formula (4). Particularly, the initial resource of item v_1 is distributed to users u_1, u_2 and u_4 according to the transfer rate of each link. The resource allocation of u_1, u_2 and u_4 from item v_1 is 0.48, 0.52 and 0.4 respectively. After the other items finish resource distribution, the resource allocation of every user can be got. For example, 0.48 in total for user u_1 with resource 0.48 transferred from item v_1 and 0 from v_2 respectively.

Similarly, in the second round, resource transfers from user nodes back to item nodes according to the ratio of the rating each item got to the sum of all ratings by this user. Nonetheless, no normalization needs to be done in the round, as these ratings are all given by the same user. Hence, we can derive the final resource allocation to an item v_j by

$$R_{\text{fin}}(v_j) = \sum_{u_i} R_{\text{fin}}(u_i, v_j),$$

where $R_{\text{fin}}(u_i, v_j)$ denotes the final resource from user u_i to item v_j, which is

$$R_{\text{fin}}(u_i, v_j) = \frac{r(u_i, v_j)}{\sum_{v_j} r(u_i, v_j)} \times R(u_i). \tag{5}$$

Example 6: Recall Example 4. In Fig. 2(c), the numbers around the edges are the transfer rates from user nodes to item nodes. The resource is distributed from user nodes to item nodes in the second round by formula (5). After every user distributes the resource to the items with which there is a link, the final resource for every item can be summed up. For example, the final resource is 0.19 for item v_2 and 0.05 for item v_4. So, in this bipartite network sample, the recommendation result for user u_2 in our proposed algorithm is item v_2, and this is different from the result in the baseline algorithm which is item v_4 with a resource value of 0.3.

The correctness of the method remains, since the different weighting of initial resource allocation and distribution only affects the volume of transferred resource but not the distribution process. Furthermore, this rating-aware resource distribution allocates the user taste resource discriminatively. Therefore, the final resource allocation is expected to provide a more accurate suggestion on the top-n recommendation. We verify the effectiveness of the proposed algorithm in Section 4.

4   Experiments

This section presents the experimental results and analyses.

4.1   Experiment Setup

We conducted experiments on a real dataset **MovieLens** (http://www.grouplens.org), which is one of the most famous datasets for evaluating personal recommendation. It consists of 943 users, 1682 movies and about 100,000 rating records. Users rate movies according to their interest with discrete numerals from 1 to 5. The data is cleaned by removing users who rated less than 50 movies. The sparsity of the dataset is about 94.6%, meaning that the ratings by users to movies are rather insufficient. Further, we select the records if the rating exceeding the average rating of the user as primary data set, in which there is about 54800 ratings in number. Then we pick out ratings from the primary data set randomly by a ratio of 20% to construct the test set with a scale of 10960 records, which would be done five times for cross-validation. The rest ratings in the primary data set and the others records which are not included in the primary data set are the members of our training set with a scale of 83640 rating records.

For top-n recommendation, *hit ratio* and *average rank score* are two popular evaluation metrics while comparing different algorithms.

- Hit ratio counts the number that movies of the test set occur in the recommendation list and then uses the total length of recommendation list as the divisor to get the final ratio value, i.e.,

$$hr = \frac{\sum_{i \in M} \sum_{j \in N} \sum_{k \in T} H_i(j,k)}{M * N},$$

  where M denotes the number of users, N denotes the number of items recommended for each user, T denotes the number of items in test set for each user, and $H_i(j,k)$ is an indicator function,

$$H_i(j,k) = \begin{cases} 1, & rl(i,j) = tl(i,k) \\ 0, & \text{otherwise} \end{cases},$$

  where $rl(i,j)$ is the j-th item for user i in the recommendation list, and $tl(i,k)$ is the k-th item of user i in the test set.

- Average rank score measures the position of each movie of the test set in the sequence of all unrated movies, and smaller score value means better recommendation result. The average rank score is the average of these scores for each record in the test set which is shown as below:

$$r_{ik} = \frac{loc(tl(i,k))}{Q_i'},$$

here $loc(tl(i,k))$ is the location of item $tl(i,k)$ in the sequence for user i in the test set, and Q_i' is the number of unrated items by user i in the sequence.

4.2   Comparing with The Baseline Algorithm on Hit Ratio

In this set of experiments, we evaluate the effects of Average-Rating Criterion and Taste with Dislike (labeled by "ARC" and "TD", respectively) in the user taste model, Rating-Aware Initial Configuration and Rating-Aware Transfer (labeled by "RAIC" and "RAT", respectively) in the resource allocation.

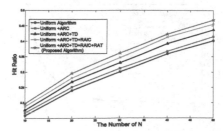

Fig. 3. Effect of our user taste model and resource allocation method

As Fig.3 shows, the user-specific average ratings criterion outperforms the median one when specifying the taste of users. Median method ignores the difference of rating habits between different users while average rating method judges whether users like an item according to more particular and personal criterions.

When the tastes of users are specified as dislike together with like, higher hit ratio was achieved than using only like information as user tastes. The dislike information as negative resource can spread over the bipartite network to rectify the single taste resource allocation. A comprehensive taste considering both like and dislike of users got a more accuracy recommendation result.

The initial configuration of resource based on comprehensive user taste model sets the values of items which had been rated by current user as 1 or -1 representing like or dislike. Considering the extent of a user likes or dislikes these items could still be very different, the ratings by users to items are utilized to weight the initial resource. This is the rating-aware initial configuration. We can tell from Fig.3 that rating-aware initial resource configuration got better result.

The distribution based on degree considered only the relationship of user and item nodes, but the weight for each link between item and user is not measured. The rating-aware transfer uses the ratio of the rating by a user to an item to the sum of ratings by the particular user to all items or all users to the given item as the importance of the edge between this user and the item. The rating-aware transfer weights degree-based transfer with the importance of links, so its recommendation accuracy is higher.

The proposed algorithm emphasizes the taste of dislike as much as the taste of like which are specified according to the average rating of each user; the resource allocation and distribution are adjusted to consider the rating ratios and the transfer rates of the edges. Compared with the uniform algorithm, which is also named Network Based Inference (NBI), our algorithm performs obviously better with higher hit ratio.

4.3   Comparing with Baseline Algorithm on Average Rank Score

We compare the NBI and our proposed algorithm with average ranking score metric. Avg-Rating, Com-Taste, RA-Initial and RA-Transfer denote the average rating based coarse-graining, the comprehensive taste with both like and dislike, rating-aware initial configuration and rating-aware transfer of resource respectively. They are used to indicate the algorithm adopted the improvements that they represent and the ones before them. $< r >$ denotes the average rank score.

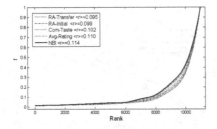

Fig. 4. Comparison on average rank score

As Fig. 4 shows, *rank* is the total scale of items in the test set which are used to validate the recommendation results while the sequence attribute is ignored temporarily, $<r>$ is the rank score. We can see that the rank scores for the five curves increase exponentially while the length of rank grows, especially start around 6000. Averaged over records in the test set, the average rank scores for the five implementations are shown respectively in the legend. It is easy to find that all the improvements are effective, and the introducing of dislike as taste got the biggest improvement. Our proposed algorithm achieves better performance than NBI by 16.7% using average rank score metric, which is accordant with the conclusion using hit ratio metric.

4.4 Comparing with Other Three Variants

We compare the proposed algorithm with an algorithm based on mass diffusion and two popular variants of bipartite network projection to further validate effectiveness with both hit ratio and average rank score metrics.

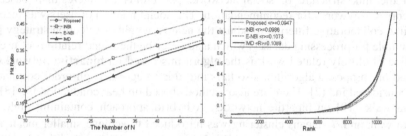

Fig. 5. Comparisons with other three algorithms

- **IMD** [11] is an improvement of Mass Diffusion which takes into account the average degree of user nodes to weight the initial resource distribution;
- **E-NBI** [12] is a variant of uniform NBI which depresses the impact of high-degree with a negative exponential function to improve the accuracy;
- **INBI** [13] combines weighted bipartite network with a tunable parameter to depressing high-degree nodes for top-n recommendation.

In Fig.5, the left part demonstrates the comparison with hit ratio metric and while the recommend list varying from 10 to 50 the proposed algorithm achieves the best result all the time; the right part is the result of comparison with average rank score, the proposed algorithm outperforms the other ones by 13.0%, 11.7% and 3.9% on average, respectively.

5   Related Work

Personal recommendation has been a topic that draws much attention.

Content-Based Method extracts features for items, and a profile is created for a user with the features, which is then utilized to find out the most similar items to those collected. However, feature extraction is not easy, and the recommendation is restricted to items very similar to those have been liked [1]. Semantic reasoning was used to overcome the shortage of vector space model [14]. This kind of algorithms needs to describe item content and extract properties of rated items to build user profiles, so the cold start problem is serious.

Contrarily, **collaborative filtering** does not rely on item content. It utilizes rating behavior data to select the most similar neighbor, and makes recommendation on the assumption that who behaved similarly in the past will behave similarly in future. The key steps of collaborative filtering are computing the similarity and generating the recommendation from the nearest neighbors. Representative work includes [2-4]. Due to the vast quantity of items and users but quite finite ratings, the sparsity problem turns into the bottleneck to raise the effectiveness.

Matrix factorization [5, 6] characterizes both items and users by vectors of factors inferred from rating patterns, and high correspondence between item and user factors leads to a recommendation. The single value decomposition [15] is the most commonly used method to reduce the matrix dimension. The complexity of the computation is so high that few practice applications adopt this kind of algorithms.

Social-based method recommenders the user with what his neighbors like according to the link structure of social network. [7] builds an interestmap based on co-occurring keywords for recommendation. The social relation of users was incorporated into collaborative filtering to adjust the nearest neighbor selection strategy [16]. It is suitable for processing datasets which also describe the friend relations of users.

A line of closely related work is the algorithms based on **bipartite networks** [8-10], and our proposed algorithm also falls into this category. High-order correlations were considered in [17]. There are also methods based on heat conduction [9, 18] and random walk [19] on bipartite networks. A hybrid approach containing multi-step random walk and k-means clustering was introduced to achieve smaller mean absolute error and root mean square error [20]. It is more precise than traditional collaborative filtering recommendation algorithms, yet its accuracy and efficiency could be further improved.

6   Conclusion

In this paper, we have proposed a negative-aware and rating-integrated personal recommendation algorithm based on bipartite network projection. It takes the advantages of the information implied by user ratings to differentiate dislike cases of user tastes, and weight the resource distribution process. Better empirical results on real data are obtained regarding both hit ratio and average rank score.

Acknowledgement. This research is partially supported by the National Natural Science Foundation of China under Grant NO.61302144 and 62303062.

References

1. Marko, B., Yoav, S.: Fab: Content-based Collaborative Recommendation. Comm. of the ACM 40(3), 66–72 (1997)
2. Konstan, J.A., Miller, B.N., Maltz, D., Herlocker, J.L., Gordon, L.R.: GroupLens: Applying Collaborative Filtering to Usenet News. Comm. of the ACM 40(3), 77–87 (1997)

3. Khoshneshin, M., Street, N.W.: Collaborative Filtering via Euclidean Embedding. In: ACM RecSys 2010, Barcelona, Spain, September 26-30 (2010)
4. Koren, Y.: Collaborative Filtering with Temporal Dynamics. In: ACM KDD 2009, Paris, France, June 28-July 1 (2009)
5. Koren, Y., Bell, R., Volinsky, C.: Matrix Factorization Techniques for Recommender Systems. IEEE Computer 42(8), 30–37 (2009)
6. Shi, Y., Larson, M., Hanjalic, A.: Mining Contextual Movie Similarity with Matrix Factorization for Context-Aware Recommendation. ACM TIST 04(01), 1601–1619 (2013)
7. Liu, H., Maes, P.: InterestMap: Harvesting Social Network Profiles for Recommendations. In: IUI 2005, San Diego, California, USA (January 9, 2005)
8. Zhou, T., Ren, J., Medo, M., Zhang, Y.: Bipartite Network Projection and Personal Recommendation. Phys. Rev. E 76(4), 46115 (2007)
9. Zhang, Y., Blattner, M., Yu, Y.: Heat Conduction Process on Community Networks as a Recommendation Model. Phys. Rev. Lett. 99, 154301 (2007)
10. Zhang, Y., Medo, M., Ren, J., Zhou, T., Li, T., Yang, F.: Recommendation Model Based on Opinion Diffusion, Europhys. Europhys. Lett. 80(2008), 68003 (2008)
11. Liu, J., Zhou, T., Wang, B., Zhang, Y.: Effects of User's Tastes on Personalized Recommendation. Int. J. Mod. Phys. C 20, 1925–1932 (2009)
12. Zhou, T., Jiang, L., Su, R., Zhang, Y.: Effect of Initial Configuration on Network-Based Recommendation. Europhys. Lett. 81(2008), 58004 (2008)
13. Xia, J., Wu, F., Xie, C., Tu, J.: INBI: An Improved Network-Based Inference Recommendation Algorithm, In: IEEE NAS 2012, June 28-30 (2012)
14. Fernandez, Y.B., Arias, J.P., Solla, A.G., Cabrer, M.R., Nores, M.L.: Providing Entertainment by Content-based Filtering and Semantic Reasoning in Intelligent Recommender Systems. IEEE Tconsum. Electr. 54(2), 727–735 (2008)
15. Sarwar, B., Karypis, G., Konstan, J., Riedl, J.: Incremental Singular Value Decomposition Algorithms for Highly Scalable Recommender Systems. In: Proceedings of ICCIT, April 2-4 (2002)
16. Liu, F., Lee, H.J.: Use of Social Network Information to Enhance Collaborative Filtering Performance. Expert Syst. Appl. 37(7), 4772–4778 (2010)
17. Liu, J., Zhou, T., Che, H., Wang, B., Zhang, Y.: Effects of High-order Correlations on Personalized Recommendations for Bipartite Networks. Physica A 389(2010), 881–886 (2010)
18. Liu, J., Zhou, T., Guo, Q.: Information Filtering via Biased Heat Conduction. Phys. Rev. E 84, 37101 (2011)
19. Quan, J., Fu, Y.: A Novel Collaborative Filtering Algorithm Based on Bipartite Network Projection. JDCTA 6(1), 391–397 (2012)
20. Sawant, S.: Collaborative Filtering using Weighted Bipartite Graph Projection: A Recommendation System for Yelp. In: CS224W: Social and Information Network Analysis (December 10, 2013)

Sentiment Analysis on Twitter through Topic-Based Lexicon Expansion

Zhixin Zhou, Xiuzhen Zhang, and Mark Sanderson

Department of Computer Science and IT,
RMIT University, Melbourne, VIC 3000
{zhixin.zhou,xiuzhen.zhang,mark.sanderson}@rmit.edu.au
http://www.rmit.edu.au

Abstract. Supervised learning approaches are domain-dependent and it is costly to obtain labeled training data from different domains. Lexicon-based approaches enjoy stable performance across domains, but often cannot capture domain-dependent features. It is also hard for lexicon-based classifiers to identify the polarities of abbreviations and misspellings, which are common in short informal social text but usually not found in general sentiment lexicons. We propose to overcome this limitation by expanding a general lexicon with domain-dependent opinion words as well as abbreviations and informal opinion expressions. The expanded terms are automatically selected based on their mutual information with emoticons. As there is an abundant amount of emoticon-bearing tweets on Twitter, our approach provides a way to do domain-dependent sentiment analysis without the cost of data annotation. We show that our technique leads to statistically significant improvements in classification accuracies across 56 topics with a state-of-the-art lexicon-based classifier. We also present the expanded terms, and show the most representative opinion expressions obtained from co-occurrence with emoticons.

1 Introduction

Both machine-learning and lexicon-based approaches have been adopted to do sentiment analysis on Twitter. Machine-learning approaches to sentiment analysis usually require annotated text and are known to be domain-dependant. Annotation is generally costly to obtain, but lack of labeled data in the target domain can lead to deteriorated classification performance. It has also been shown in existing work [14] that better performance is achieved when using all words as features. As such, the feature list of a supervised classifier often contains spurious patterns which are difficult to make sense of by a human reader. As has been indicated by Thewall et.al [17], supervised classifiers may harness nonsentiment features and falsely identify sentiment.

Lexicon-based approaches relies on opinion lexicons to classify text. Words from such lexicons are direct indicators of sentiment and transparent to human readers, and the polarities of most opinion words are not domain-dependent. As such, these approaches can achieve stable performance across domains.

H. Wang and M.A. Sharaf (Eds.): ADC 2014, LNCS 8506, pp. 98–109, 2014.

However, for more accurate classification it is desirable to capture contextual polarities of words [3], especially so when dealing with short social text such as Twitter tweets, where single-sentence status updates are common and the number of features are rather limited. For example, the word "big" implies positive emotion in "My new office is BIG!!" but negative emotion in "it's too big to fit into your pocket". Another challenge for lexicon-based approaches on short social text is to identify the polarity of informal expressions of sentiments, such as abbreviations and misspellings. In a tweet positive expressions may include "+1", "hear hear". A negative expression of opinion can be "O come on", "lmao" (laugh my a** off). Such informal expressions may evolve and emerge over time, and to the best of our knowledge, no existing opinion word lexicon provides sufficient coverage of these informal expressions.

In addition to opinion lexicons, emotions can also be expressed through emoticons. Unlike literal words, emoticons usually have a stable polarity across domains and have been widely used for sentiment classification. In fact, the default twitter search allows users to add emoticons to the query to find positive or negative tweets, and the returned results usually contains emoticons. However, the majority of tweets do not have emoticons. Our statistics show that only 9.40% (7.37% positive and 2.03% negative) of the tweets in the Microblog Track 2011 collection have at least one emoticon [1].

Insufficient lexicon coverage and the limitations of using emoticons have motivated this study. We propose an automatic lexicon expansion technique to improve the coverage of the sentiment lexicon employed by the classifier, by measuring the mutual information between potentially sentiment-bearing words and the emoticons. We specifically study the following research questions,

1. Can emoticon-aided lexicon expansion improve the performance of a sentiment classifier?
2. Can topic-biased emoticon-aided lexicon expansion improve the performance of sentiment classification?

Our expansion technique is based on a simple intuition. For tweet that looks like "+1 :)", or "hear hear :P", we may use the polarity implied by the emoticons ":)" and ":P" to infer that "+1" and "hear hear" are positive. Specifically, we use the point-wise mutual information (PMI) between each word (or symbol) and a known set of emoticons (see Table 1) to measure the sentiment polarity of the word. Our technique differs from Turney et.al [20] not only in that we use positive emoticon groups and negative emoticon groups as references for PMI calculation, but also in the way we deal with negation. In their study, negations in the text are not handled. As such, *"the online service was not excellent at all"* would be treated as evidence that *online service* is a positive phrase. In our appoach we apply negation detection mechanism to flip such cases so that *"I don't like their online service :("* would be counted as a co-occurrence of *like* and *:)*.

---

[1] We matched all tweets against the emoticons from Table 1.

To answer the second research question, we use hashtags to create a collection of tweets from 56 topics. Hashtags are a type of metadata used in Twitter community to add additional context to tweets, by prefixing a word with a hash symbol, such as #twitter. Tweets with the same hashtag can often be considered to be about the same coarse "topic", though in many cases not a topic in the common sense. For instance, hashtags such as #justsaying and #nowplaying are not typical topics. Nonetheless, a hashtag groups tweets of a similar concern, thus enforcing a coarse semantic relation between the tweets. Intuitively, topic-based lexicon expansion is more difficult when such semantic relations are weaker. Therefore, evaluation conducted on this hashtag-based topic collections should be more rigorous than one based on a human-labeled topic collection. We show that per-topic expansion leads to significantly better performance than global expansion done on the combined set of tweets from all topics.

2   Related Work

Sentiment classification models on Twitter can be broadly categorised as machine-learning and lexicon-based approaches, though some algorithms [22] have elements of both. Machine-learning approaches typically requires labeled training data, and often use text features as well as emoticons as features to train the model. To reduce the cost of obtaining labeled training data, some sentiment-bearing tokens (e.g. emoticons, sometimes even hashtags) with known sentiment polarities have been used to automatically collect training instances. Go et.al [8] use emoticons as "noisy labels" to obtain instances, and classify with various classifiers, among which MaxEnt has achieved an accuracy of 83% on their test set. Pak et.al [12] also use emoticons as labels for training data and has built a multinomial Naive Bayes classifier based on N-gram and POS-tags as features. Davidov et.al [6] utilize 50 Twitter tags and 15 smileys as sentiment labels. Liu et.al [9] utilize both manually labeled data and noisy labeled data for training, where emoticons are used to smoothen a supervised classification model.

Recently, lexicon-based approaches [16,18] have gained popularity. Lexicon-based approaches can achieve stable performance across domains, and the features used are more transparent to a human user. They typically employ opinion word lexicons, such as SentiWordNet [1] and General Inquirer lexicon [5], to match against the text to be classified. The presence of annotated (typically with opinion scores or scales) lexical items (opinion words/phrases) are processed with linguistic rules to compute an overall semantic orientation of the document.When such a pre-defined lexicon is not present, other unsupervised methods [20] can be adopted to automatically construct a lexicon.

Various studies aim to solve the problem of adapting a machine-learning model for classification in a new domain. Such techniques are often referred to as transfering learning techniques[2], and the majority [21,7,4] of them require at least a small amount of labeled data from the target domain. The approaches that

---

[2] The terminologies *domain adaptation* and *transfer learning* are often used interchangeably, and in this study we stick to the latter.

do not require labeled data in the target domain include the Structural Correspondence Learning (SCL) algorithm by Blitzer et.al [3] and the dimensionality reduction approach by Pan et.al [13]. These approaches generally utilize the common features between the source domain and the new domain to establish a link for knowledge transfer. In [3] the link is built through the pivot features and in [13] the link is built through a latent space that minimizes the difference between different domain distributions.

To do domain-specific classification with lexicon-based classifiers several approaches have been developed. Ponomareva et.al [15] use a graph to model a group of labeled and unlabeled documents, and update the sentiment scores of unlabeled documents based on nearest documents. This approach operates at document level, therefore the classification process is no longer transparent to a human user. Domain-specific lexicon expansion is another way to adapt a lexicon-based classifier to different domains. The approach by Thelwall et.al [17] requires human intervention to annotate the corpus and do a small amount of term selection for different domains. Choi et.al [5] adapt an existing lexicon to a given domain through an optimization framework, where phrase-level subjectivity annotation is required. Turney et.al used two reference words, *excellent* and *poor*, to represent the two extremes of opinions, and use pointwise mutual information (PMI) to calculate the semantic orientation of words. Becker's work [2] is the most related to ours. Their study is also built around Turney's idea [20] of using PMI to construct a lexicon, but instead of computing PMI between a word and *reference words*, they compute the PMI between a word and *sentiments*. The sentiments are tagged by a polarity classifier before PMI calculation.

While emoticons have been used in sentiment analysis on twitter [8,12,6,9], and PMI calculation has been used to construct opinion word lexicons [20,2], no existing work uses emoticon as reference tokens to do domain-specific PMI-based expansion. Also, when computing PMI, existing studies do not consider negation handling, therefore the cooccurrence of a negated word/phrase and a reference word [20] (or sentiments [2]) will give misleading information on the polarity of the word.

3   Emoticon-Based Sentiment Lexicon Expansion

3.1   Classification Framework

While we focus on lexicon-based classifiers in this study due to the stability of their performance across domains and the transparency of the classification process, the classification results of a machine learning approach is also presented as a reference.

SentiStrength [18] (SS) is a state-of-the-art lexicon-based classifier. Incorporating a booster word list, an emoticon list, an idiom list, a negation word list, a question word list, a slang list and a general opinion word list, it further applies linguistic rules to compute the overal sentiment polarity. The opinion word scores are integers ranging from [-5, -1] for negative words and [1,5] for positive words, where -1 and 1 denotes neutral words. The core of the SentiStrength lexicon is a general opinion word list of 298 positive and 465 negative terms, some of which include wild cards. For example, *abandon** would match all words that starts with *abandon*. For details of its algorithm please refer to [19]. SentiStrength is designed to report binary (positive or negative), trinary (positive, negative or neutral) and single scale(-4 to +4) results. We use its binary output, as our collection comprise of positive and negative tweets. Before expanding the SentiStrength lexicon, linear scaling has been performed to transform the semantic orientation score calculated from PMI to the SentiStrength scales. We merge our expanded terms with the SentiStrength terms without further modification. We disabled the emoticon word list in SentiStrength, since we are using the emoticons as class labels. In our pilot study, SentiStrength has achieved 90.32% when using emoticons, but its performance dropped to 76.92% when the emoticon list was disabled.

The Naive Bayes Multinomial (NBM) classifier uses all words as features without stemming, nor normalization to lower case. Naive Bayes has achieved superior performance at a per-topic average of 93.15%. Despite the high accuracy, the features used are hard to make sense of. In our experiments, NBM is confined to use only opinion words as features to make a relative fair comparison. The original feature list includes the words from SentiWordnet [1], which comprises of 21109 opinion words, many times larger than the opinion lexicon used by SentiStrength.

3.2   Lexicon Expansion

For lexicon-based classification on Twitter, a widely acknowledged problem is the word mismatch between tweet content and general opinion word lexicons. For example, words with repeated letters are commonly seen in tweets, such as *huuuungry*. Some studies [8] replace all repeated letters with two repeated letters in each word, and leave it to the classifier to leverage these features. While being effective in this particular case, this approach is insufficient to tackle many other forms of informal spellings, such as *gr8t*. In fact, these informal expressions (including informal spellings and abbreviations) are evolving over time, making it difficult for any rule-based approach to adapt to the changes.

In this study, we propose to use mutual information between sentiment tokens [3] and emoticons to adjust the sentiment strength scores of words in a general sentiment lexicon and add unseen variations of sentiment words into the lexicon. There are an abundant amount of emoticon-bearing tweets on Twitter, forming the basis of our technique. With our expansion technique, the polarities of the informal expressions as well as the less common emoticons can be automatically computed.

---

[3] A token can be a word, an abbreviation, an emoticon not found in our emoticon list shown in Table 1, or any other text segment potentially bearing sentiment.

Table 1. List of Emoticons used (Manually selected from http://en.wikipedia.org/wiki/List_of_emoticons, based on frequency and clarity)

Emoticon	Polarity	Emoticon	Polarity	Emoticon	Polarity	Emoticon	Polarity
:)	+	:-)	+	;(	-	:/	-
:>	+	;)	+	:(	-	:-(	-
;-)	+	;>	+	:[	-	:<	-
B-)	+	8-)	+	8-(	-	:-o	-
B->	+	8->	+	:-&	-	:/)	-
:->	+	:-)))	+	:-c	-	:-C	-
:D	+	:-D	+	:-<	-	;-C	-
:-P	+	^_^	+	:-|	-	:'-(	-
^.^	+	^_^'	+	:~-(	-	-_-	-
D=	+	:p	+	>_<	-	=_=	-
;D	+						

The following formula is used to calculate the point-wise mutual information (PMI) between any two words,

$$PMI(word_1, word_2) = log_2\left(\frac{p(word_1, word_2)}{p(word_1)p(word_2)}\right) \quad (1)$$

The semantic orientation (SO) of a word is given by Equation 2,

$$SO(token) = PMI(token, +ve) - PMI(token, -ve) \quad (2)$$

where $+ve$ and $-ve$ represents positive and negative emoticons respectively.

Our lexicon expansion technique is based upon the following assumption: The sentiment orientations of emoticons such as ":)" and ":(" are relatively stable across all tweets. Therefore, positive and negative emoticons can then be used to represent the two extremes of opinions. Turney's approach also implicitly assumes that words in the context of the reference word tend to share its polarity. This second assumption is not valid in sentences with negations, for example, "I don't like this guy:(". As such, in tweets with negation words, we flip the polarity of the emoticon before calculating mutual information. Also, we map all negative emoticons to ":(" and all positive emoticons to ":)". The sentiment orientation (SO) of any sentiment token is then calculated as,

$$SO(token) = log_2\left(\frac{hits(token, +ve) \times hits(-ve)}{hits(token, -ve) \times hits(+ve)}\right) \quad (3)$$

In Equation 3, $hits(+ve)$ and $hits(-ve)$ are global counts of positive and negative emoticons respectively. $hits(token, +ve)$ represents the number of co-occurrences of the token and positive emoticons, or the token and negative emoticons in the presence of negation. In the above-mentioned example, the co-occurrence of *like* and *:(* is counted as a hit between *like* and *:)* due to the presence of the negation word *don't*. The list of regular expression patterns for negation detection are shown in Table 2.

Table 2. Stats on negation patterns collected on the whole Microblog Track collection. Column Frequency shows the percentage of tweets in which the pattern has a match.

Pattern	Example	Frequency
[Nn] [Oo] [Tt]	not	3.83%
[Dd] [Oo] [Nn] '*[Oo]*[Tt]+	don't	3.10%
[Cc] [Aa] [Nn]+'*[Oo]*[Tt]+	can't	1.70%
[Aa] [Ii] [Nn] '*[Tt]	ain't	0.97%
[Ii] [Ss] [Nn] '*[Oo]*[Tt]	isn't	0.30%
[Hh] [Aa] [Vv] [Ee] [Nn] '*[Oo]*[Tt]	haven't	0.29%
[Ww] [Oo] [Uu] [Ll] [Dd] [Nn] '*[Tt]	wouldn't	0.20%
[Cc] [Oo] [Uu] [Ll] [Dd] [Nn] '*[Tt]	couldn't	0.16%
[Hh] [Aa] [Ss] [Nn] '*[Oo]*[Tt]	hasn't	0.05%
[Bb] [Aa] [Rr] [Ee] [Ll] [Yy]	barely	0.03%
[Hh] [Aa] [Rr] [Dd] [Ll] [Yy]	hardly	0.02%
[Hh] [Aa] [Dd] [Nn] '*[Oo]*[Tt]	hadn't	0.02%
[Ww] [Uu] [Dd] [Nn] [Tt]	wudn't	0.00%
[Cc] [Uu] [Dd] [Nn] [Tt]	cudn't	0.00%

Not all words bear sentiments. We use the part-of-speech tags of words to filter out potential opinion words and only include nouns, adjectives, verbs, adverbs, abbreviations emoticons and interjections. Part-of-speech tagging is done with the TwitterNLP package [11]. This set of tags was chosen after experimenting with differnt combinations of tag sets in preliminary experiments, and using this set to do classification has led to high classification accuracy, while still keeping meaningful words.

4   Experiment

4.1   Dataset   ,

Our experiments are based on a 56-topic collection of tweets from the TREC Microblog Track 2011 and the Stanford Sentiment140 collection. This dataset is generated by aggregating opinion-bearing tweets with 56 popular hashtags in the combined collection. The tweets do not come with labels – we use the emoticons as their labels, as has been done in previous studies. In our experiments on this dataset, 10-fold cross validation is carried out and the lexicon expansion was done on the training set only. The emoticons used for lexicon expansion are not used as features by any classifier in the classification process. The collection is generated via three steps,

1. Merge all tweets from TREC Microblog Track 2011 collection [10] and the Stanford Sentiment 140 collection [8].
2. Filter out all tweets that do not bear any emoticon from Table 1.
3. Group the remaining tweets by hashtag and keep the groups with at least 100 tweets.

The whole collection includes 16683 positive and 6099 negative tweets. Full details of the collection are shown in Table 3.

Table 3. Statistics of the 56-topic collection

Topic	# Pos	# Neg	Topic	# Pos	# Neg	Topic	# Pos	# Neg
glee	99	17	follow	255	14	nowfollowing	165	0
2	214	152	followfriday	2529	143	agoodboyfriend	119	6
bgt	133	114	bieberd3d	484	166	neversaynever	484	74
e3	119	136	asot400	190	135	nowwatching	131	5
fb	934	1101	februarywish	228	25	oneofmyfollowers	109	51
1	425	187	followback	111	3	iranelection	79	432
f1	122	152	iphone	96	236	icantdateyou	93	79
humor	143	1	improudtosay	395	16	idontunderstandwhy	48	82
fml	10	99	iremember	153	150	jedwardlipstick	125	16
jfb	379	9	justsaying	103	13	marsiscoming	205	54
music	162	15	musicmonday	383	38	myweakness	217	52
bsb	131	164	nowplaying	590	70	neversaynever3d	456	37
fail	72	400	random	90	30	spymaster	99	70
ff	2098	130	purpleglasses	189	61	questionsidontlike	16	89
np	606	72	seb-day	428	71	shoutout	464	29
nw	139	12	squarespace	282	656	superbowl	233	58
tcot	140	34	twitteroff	152	22	teamfollowback	499	22
trackle	16	145	twitition	193	2	twitter	111	75
tfb	117	2	wheniwaslittle	120	75			

4.2   General Lexicon vs. Global Expansion

In this experiment we aim to answer the first research question by contrasting the performance of SentiStrength (SS) and Naive Bayes Multinomial (NBM) before and after doing global expansion (GE). With GE, tweets from all topics are merged into a single collection, on which lexicon expansion is based upon. The expanded lexicon is then used to do per-topic classification. For NBM, words from SentiWordNet [1] is used as the original feature list, and SS uses its own lexicon. After expansion, the expanded terms are added to the feature list of NBM, and also added to the lexicon of SS.

The classification results are shown in Table 4. Paired t-test was done across the 56 topics to show statistically significant improvement with SentiStrength, with a p value of 1.6e-5. This indicates that PMI-based lexicon expansion does indeed lead to a better lexicon for classification with SentiStrength. With NBM however, the classification performance after global lexicon expansion is higher but not statistically significant.

4.3   Topic-Based Expansion vs. Global Expansion

We contrast the performance of SS and NBM with Global Expansion (GE) against Topic-based Expansion (TE) to answer the second research question. As is shown in Table 4, topic-based expansion has indeed lead to significantly better

Table 4. Classification with no lexicon expansion, Global Expansion and Topic-based Expansion. Columns *All, +* and *-* shows the overall accuracy, positive precision and negative precision respectively. NBM refers to Naive Bayes Multinomial, and SS refers to SentiStrength.

Classifier	No Expansion (%)			With GE (%)			With TE (%)		
	All	+	-	All	+	-	All	+	-
NBM	82.38	83.77	38.62	82.61	83.43	41.50	82.72	83.62	40.02
SS	76.92	84.33	44.77	**79.99**	**87.51**	**48.99**	**85.24**	**88.29**	**58.03**

classifcation resulst with SentiStrength. Paired t-test done across 56 topics shows significant improvement with a p value of 3.97e-9. In fact, the performance of SentiStrength with topic-based expansion has even exceeded NBM with statistical significance (p value = 1.69 e-45 against NBM with GE and 2.89 e-45 agaist NBM with TE). Note though, NBM and SS are not using the same feature list. NBM is using an opinion word lexicon of more than 21109 words while the size of the SS's core lexicon is only around 2000 words.

Figure 1 shows the similarity between expanded terms from different topics. Cosine similarity was measured between each pair of topics to compare how similar the 50 most opinionated terms are. The subjectivity of a term is evaluated by PMI. As is shown in the heatmap, the terms are quite different. The highest similarity was found between *justsaying* and *agoodboyfriend*, with a similarity score of only 0.47, which indicates highly diverged extensions to the original lexicon.

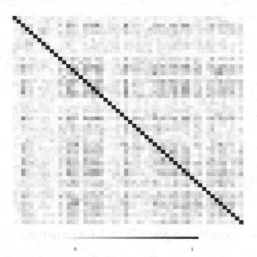

Fig. 1. Cosine similarity between the expanded terms (top 50) from each topic. The topics on the X axis (from left to right) and Y axis (from top to bottom) are shown in the same order as in Table 3.

Table 5. The 10 topics that benefit most and the 10 topics that suffer most from lexicon expansion

Topic	Improvement	#pos	#neg	Topic	Improvement	#pos	#neg
questionsidontlike	+88.97%	16	89	music	-14.07%	165	15
fml	+42.77%	10	99	follow	-9.94%	255	14
iphone	+40.78%	96	236	humor	-7.69%	143	1
trackle	+36.12%	16	145	musicmonday	-3.62%	383	38
e3	+34.34%	119	136	tfb	-3.40%	117	2
fail	+31.72%	72	400	neversaynever3d	-2.69%	484	74
spymaster	+29.51%	99	70	idontunderstandwhy	-2.17%	48	82
squarespace	+26.06%	282	656	seb-day	-1.04%	428	71
tcot	+26.06%	140	34	random	-0.71%	90	30
justsaying	+24.25%	103	13	nowfollowing	-0.45%	165	0

Topics on which classification performance have been affected most by topic-based expansion, both negatively and positively, are shown in Table 5. Abbreviations like *fml* (f*** my life), *tcot* (The changing of times), *tfb* (teamfollowback are not self-explanatory. Both conventional topics such as *iphone* and *e3*, and hashbag-based topics such as *questionsidontlike* and *fml* benefit hugely from topic-based lexicon expansion. Among the topics that suffer most, both conventional topics (*music*) and hashtag-based topics (*tfb*) are found too.

Table 6 provides two samples of lexicon expansions. As is shown in the table, our expansion technique can effectively identify the polarities of 1) abbreviations. For instance, *lmao*(laugh my a** off), *lmfao* (laugh my f***ing a** off), *smh* (shaking my head) are associated with scores of the proper polarities. 2) informal words such as yup, ugh. 3) topic-specific words including *gay* from topic *icantdateyou* and *making, want* from topic *iphone*. When mentioning *gay* in topic *icantdateyou*, people are often making jokes, for example: *@MissSarahDan hahahahaha that's exactly what I was thinking :)* or *#icantdateyou cause you're gay! Lol XxX*. The word *making* would be considered neutral in most cases, but is often mentioned in the contexts such as *making it hard to look at any screen* and *why are you making me wait until noon for #iPhone OS 3.0?* in the topic *iphone*, thus having a strongly negative PMI score. Similarly, the word *want* is frequently used in the pattern of *I want ..., but ...*, showing a need that has not been satisfied. As such, the seemingly neutral word has been assigned a strongly negative score, becoming a useful feature in the context of topic *iphone*.

Finally, we contrast our negation-aware PMI expansion to traditional PMI expansion that does not detect negations. As is shown in Table 7, no statistically significant difference have been observed between the two runs, though the average accuracy is slightly higher when using negation handling. The number of tweets per topic in this collection is limited compared to the data that a commercial company may own, which may be the reason why the effect of negation-aware PMI expansion is not clearly shown.

Table 6. Expanded term samples. Top 50 from two topics ordered by the absolute values of PMI.

Word	PMI	Word	PMI	Word	PMI	Word	PMI
expanded terms for *icantdateyou*				expanded terms for *iphone*			
love	19.27	wear	-18.83	cant	20.66	want	-20.36
dont	-18.50	smarter	-18.50	cool	20.17	seems	-19.55
haha	18.27	lmao	18.27	guess	-19.55	yup	19.43
URLSTRING	18.27	talk	-18.09	explain	19.43	worked	19.43
look	-18.09	gay	17.85	care	19.43	finish	19.43
can't	-2.24	how	-2.24	says	-19.36	didn't	-19.36
go	-2.24	am	2.09	being	-18.87	work	-18.87
don't	-1.88	have	-1.82	making	-18.87	said	-18.55
got	-1.82	call	-1.82	trying	-18.55	crashes	-18.55
date	-1.82	still	-1.82	already	-18.55	hmm	-18.55
will	1.76	want	1.76	seriously	-18.55	wants	-18.55
do	1.57	know	-1.56	keeps	-18.55	went	-18.14
being	1.35	back	1.35	gonna	-18.14	oh	-18.14
when	-1.24	even	-1.24	small	-18.14	crashing	-18.14
been	-1.24	only	-1.24	let	-18.14	cut	-18.14
wanna	-1.24	waiting	-1.24	where	-18.14	looks	-18.14
try	-1.24	play	-1.24	would	-18.14	means	-18.14
come	-1.24	see	-1.24	gone	-18.14	ugh	-18.14
smh	-1.24	be	1.18	lost	-18.14	stuck	-18.14
get	1.09	not	-1.01	feel	-18.14	omg	-18.14
are	-0.97	already	-0.82	not	6.00	can't	3.76
is	-0.77	then	0.76	don't	2.88	looking	2.88
oh	0.76	cant	0.76	happy	2.88	think	2.88
lmfao	0.76	make	0.76	find	2.88	can	2.52
aint	0.76	bigger	0.76	did	2.30	again	2.30

Table 7. PMI-based expansion with and without negation detection. Columns *All*, *+* and *-* shows the overall accuracy, positive precision and negative precision respectively.

Classifier	Handle Negation (%)			Ignore Negation (%)		
	All	+	-	All	+	-
SS	85.24	88.29	58.03	84.95	85.73	71.91

5   Conclusions

In this study we propose a structured approach to domain-dependent senti-ment analysis through lexicon expansion aided by emoticons. Our experiments have shown that emoticon-aided lexicon expansion does improve the perfor-mance of a state-of-the-art lexicon-based classifier, and topic-based expansion outperforms global expansion. This indicates that our technique is effective in domain-dependant sentiment classification.

It has been observed from the experiments that our technique is able to identify the polarities of abbreviations and informal expressions, as well as topic-specific words, making it particularly useful in classifying short informal text such as tweets. The abundant amount of emoticon-bearing tweets forms a solid basis of the application of our approach. With our approach, emoticons are only used in the expansion phase. With the better coverage of terms and expressions in social text, our approach is particularly useful in improving the classification of tweets that do not contain emoticons.

References

1. Baccianella, S., et al.: SentiWordNet 3.0: An enhanced lexical resource for sentiment analysis and opinion mining. In: LREC (2010)
2. Becker, L., et al.: AVAYA: Sentiment analysis on twitter with self-training and polarity lexicon expansion. In: SemEval (2013)
3. Blitzer, J., et al.: Biographies, bollywood, boom-boxes and blenders: Domain adaptation for sentiment classification. In: ACL (2007)
4. Bonilla, E., et al.: Multi-task gaussian process prediction (2008)
5. Choi, Y., Cardie, C.: Adapting a polarity lexicon using integer linear programming for domain-specific sentiment classification. In: EMNLP (2009)
6. Davidov, D., et al.: Enhanced sentiment learning using twitter hashtags and smileys. In: Coling 2010 (2010)
7. Davis, J., Domingos, P.: Deep transfer via second-order markov logic. In: ICML (2009)
8. Go, A., et al.: Twitter sentiment classification using distant supervision. In: CS224N Project Report, Stanford (2009)
9. Liu, K.L., et al.: Emoticon smoothed language models for twitter sentiment analysis. In: AAAI (2012)
10. Ounis, I., et al.: Overview of the trec-2011 microblog track. In: TREC 2011 (2011)
11. Owoputi, O., et al.: Improved part-of-speech tagging for online conversational text with word clusters. In: Proceedings of NAACL-HLT (2013)
12. Pak, A., Paroubek, P.: Twitter as a corpus for sentiment analysis and opinion mining. In: LREC (2010)
13. Pan, S.J., et al.: Transfer learning via dimensionality reduction. In: AAAI (2008)
14. Pang, B., et al.: Thumbs up?: sentiment classification using machine learning techniques. In: EMNLP (2002)
15. Ponomareva, N., Thelwall, M.: Do neighbours help?: An exploration of graph-based algorithms for cross-domain sentiment classification. In: Proceedings of the 2012 Joint Conference on EMNLP and CoNLL (2012)
16. Taboada, M., et al.: Lexicon-based methods for sentiment analysis. Computational linguistics (2011)
17. Thelwall, M., Buckley, K.: Topic-based sentiment analysis for the social web: The role of mood and issue-related words. JASIST (2013)
18. Thelwall, M., et al.: Sentiment strength detection for the social web. JASIST (2012)
19. Thelwall, M., et al.: Sentiment strength detection in short informal text. JASIST (2010)
20. Turney, P.D.: Thumbs up or thumbs down?: semantic orientation applied to unsupervised classification of reviews. In: ACL (2002)
21. Zhang, D., et al.: Sentiment detection with auxiliary data. Information retrieval (2012)
22. Zhang, L., et al.: Combining lexiconbased and learning-based methods for twitter sentiment analysis. HP Laboratories, Technical Report HPL-2011 (2011)

Discovering Collective Group Relationships

S. M. Masud Karim, Lin Liu, and Jiuyong Li

School of Information Technology and Mathematical Sciences,
University of South Australia, Mawson Lakes, SA 5095, Australia
masud.karim@mymail.unisa.edu.au,
{Lin.Liu,Jiuyong.Li}@unisa.edu.au

Abstract. In many real-world situations, individual components of complex systems tend to form groups to interact collectively. The grouping effectuates collective relationships. On the other hand, collective relationshsips stimulate individual components to form groups. To gain clear understanding of the structure and functioning of these systems, it is necessary to identify both group formation and collective relationships at the same time. In this paper, we define the notation of collective group relationships (*CGRs*) between two sets of individual components and propose a method to discover *CGRs* from heterogeneous datasets. The method integrates canonical correlation analysis (CCA) with graph mining to find top-k *CGRs*. Several experimental studies are conducted on both synthetic and real-world datasets to demonstrate the effectiveness and efficiency of the proposed method.

Keywords: Collective group relationships, group pair, canonical correlations, quasi-cliques.

1 Introduction

Understanding the structural and functional properties of large complex systems remains a major scientific challenge mainly due to the complicated relationships of the individual components of the systems. The individual components of many complex systems tend to form groups to interact collectively. The grouping makes the collective relationships happen. On the other hand, the collective relationships are likely to be a main force to drive individual components to form groups. The group formation and collective relationships are intertwined and govern each other. The collective relationships between groups are often more readily interpreted than those between individual components, and thus are more interesting. We call this type of relationships *collective group relationships* (*CGRs*) hereafter.

One example of complex systems illustrating *CGRs* is gene regulatory networks consisting of gene regulators and their target genes. Recently, it has been made perceptible that a disease is not simply caused by a single gene and the expression of a gene is not controlled by just one regulator. Instead, a group of gene regulators can collectively regulate a group of genes which may lead to certain diseases [21]. In this case, the collective relationship between a group of regulators and a group of genes causes the formation of the groups. Hence to gain

H. Wang and M.A. Sharaf (Eds.): ADC 2014, LNCS 8506, pp. 110–121, 2014.

insight into gene regulation and their impact on cell functions, it is essential to identify the collective relationships, and the groups of regulators and groups of genes involved in the relationships. Similarly, *CGRs* exist in many other complex systems too, such as food webs, and social collaboration networks [5].

Current research on the relationships of complex systems follows two main streams: *group discovery*, and *relationship discovery*. The former identifies groups of components of a system without considering the group relationships as the motivation of group formation, while the latter concerns with discovering and analyzing relationships between either individual components or known groups of components without caring about how the groups are formed.

One approach towards the discovery of *CGRs* can be to extend group discovery methods like clustering [9] to firstly identify 'clusters' of individual components, and then find out the cluster to cluster relationships, as *CGRs*. However, clustering separates components into 'natural' groups according to their similarity, so cluster to cluster relationships are not considered as the main factor of the clustering. Therefore, this approach is not a solution to the *CGR* problem. If we look back the example of gene regulation, although clustering methods can be used to find natural groups of co-expressed genes (i.e. genes with similar expression patterns), co-expression is not equivalent to co-regulation [24].

Alternatively, it may be possible to extend existing relationship discovery methods to reveal *CGRs*. For example, we may firstly use an existing relationship discovery method to identify pair-wise relationships to construct a *graph representation* of the relationships between individual components of a system. Then we can conduct *graph mining* to identify cliques or quasi-cliques of the graph [1,12]. Although such a graph mining approach provides qualitative or structural information of the relationships between each pair of the found groups, it cannot quantify the relationships. Another limitation of graph mining is that in some applications, the number of generated groups can be quite large, and there is no efficient method to eliminate redundant groups.

Another approach can be adopted to discover *CGRs* is to use extensions to Canonical Correlation Analysis (CCA) based methods via sparsity, commonly termed as sparse CCA (sCCA) [2,10,11,15,20,22,23]. These sCCA methods identify interacting group pairs and quantify the collective relationships between them simultaneously. The major problem associated with sCCA methods is that they only provide a quantative measure (i.e. strength) of collective relationships, without providing any qualitative or structural measure of the relationships. Some work has also been reported to incorporate the 'group effect' into classical CCA model [3,4]. However, a prior knowledge of group structure is needed. Recently Liu et al. [13] formulated group discovery as an optimization problem of CCA, and extended it to the lasso problem. However, the groups discovered by the method is too few, and the groups are disjoint while it is natural to have overlapping groups.

It is clear from above discussion that although existing methods have their own strengths, they are not suitable to address the proposed *CGR* problem alone. By looking closely at graph mining and CCA based methods (in Table 1),

Table 1. Comparison between the graph-mining and CCA based methods

Criteria	Graph Mining methods	CCA based methods
Data	Binary	**Numeric**
Measurement of relationships	**Qualitative**	**Quantitative**
Strength of relationships?	No	**Yes**
Links of relationships?	**Yes**	No
Number of groups	Too many	Too few
Nature of groups	**Overlapping**	Disjoint

it is worthy to notice that they are complementary to each other. For example, although graph mining generates no quantative measure, CCA quantifies collective interactions between groups of components. Therefore, to address the research problem, it is highly desirable to obtain a suitable integration of these two approaches (expected properties after integration are in bold-face in Table 1).

In this paper, we propose a method, *GRAPE* (**Gr**oup rel**a**tionship insca**pe**) to reveal *CGRs* from heterogeneous datasets by integrating CCA with quasi-clique mining. This integration enriches the identification of groups with both quantitative and qualitative measures of collective relationships. We evaluate the effectiveness and efficiency of *GRAPE* by applying it to synthetic and real-world datasets, and comparing it with other approaches.

2   Collective Group Relationship Discovery

2.1   Problem Statement and Definitions

Consider two sets of variables $\mathbf{X} = \{x_1, \ldots, x_p\}$ and $\mathbf{Y} = \{y_1, \ldots, y_q\}$ such that $\mathbf{X} \cap \mathbf{Y} = \phi$, representing the attributes of two different types of objects. With their given datasets, $\mathbf{D_X} = \{dx_1, \ldots, dx_p\}$ and $\mathbf{D_Y} = \{dy_1, \ldots, dy_q\}$, where dx_i and dy_j are n-vectors representing the samples of x_i and y_j ($1 \leq i \leq p$, $1 \leq j \leq q$) respectively, our goal is to identify any $\mathcal{P_X} \subseteq \mathbf{X}$ and $\mathcal{P_Y} \subseteq \mathbf{Y}$, such that $\mathcal{P_X}$ and $\mathcal{P_Y}$ are related. Hence, we define a collective group relationship (*CGR*) between $\mathcal{P_X}$ and $\mathcal{P_Y}$ as follows:

Definition 1. *A triple* $(\mathcal{P}, \mathcal{E}, \mathbf{r})$ *is called a* **CGR**, *where*

1. $\mathcal{P} = \{\mathcal{P_X}, \mathcal{P_Y} \,|\, \mathcal{P_X} \subseteq \mathbf{X}, \mathcal{P_Y} \subseteq \mathbf{Y}, \mathcal{P_X} \cap \mathcal{P_Y} = \phi\}$ *known as a group pair,*
2. $\mathcal{E} \subseteq (\mathcal{P_X} \cup \mathcal{P_Y}) \times (\mathcal{P_X} \cup \mathcal{P_Y})$ *denotes the links among the variables of* $\mathcal{P}$,
3. $\mathbf{r}$ *stands for the quantitative measure or strength of the CGR.*

Not all *CGRs* are equally interesting, specially a *CGR* with very few variables, say one variable, or a *CGR* of low strength (i.e. $\mathbf{r}$ is very small). Hence, we introduce two thresholds to refine Definition 1 to indicate the interestingness of *CGRs* as follows:

Definition 2. *For a minimum size threshold* $\theta \geq 2$ *and a strength threshold* $0 < \rho \leq 1$, *a CGR* $(\mathcal{P}, \mathcal{E}, \mathbf{r})$ *is said to be* (θ, ρ)-*associated if* $|\mathcal{P_X}| \geq \theta, |\mathcal{P_Y}| \geq \theta$ *and* $\mathbf{r} \geq \rho$.

With datasets $\mathbf{D_X}$ and $\mathbf{D_Y}$, and thresholds θ and ρ, the problem of discovering *CGRs* is to identify all (θ, ρ)-associated *CGRs* from $\mathbf{D_X}$ and $\mathbf{D_Y}$.

2.2   Algorithm for Discovering Collective Group Relationships

The proposed method *GRAPE* (**G**roup **r**elationship insc**ape**) first creates a graph representation of the relationships among the individual variables of the given datasets using correlation test, then it applies graph mining to compute all maximal quasi-cliques from the graph. Each of the quasi-cliques is split into two groups (i.e. group pair) and the strength of the collective relationship between the two groups is assessed. Finally, each of the group pairs together with its strength is validated.

As given in Algorithm 1, *GRAPE* takes data matrices $\mathbf{D_X}$ and $\mathbf{D_Y}$ as input and extracts all (θ, ρ)-associated *CGRs* from the graph representation of the individual relationships of $\mathbf{X}$ and $\mathbf{Y}$, and returns the top-k *CGRs* as output. The enrite algorithm works in three phases — (i) representation into graph, (iii) identification of *CGRs* from the graph, and (iii) selection of top-k *CGRs*.

Representation into Graph. First a graph representation of the relationships among individual variables of merged $\mathbf{X}$ and $\mathbf{Y}$ is constructed in Algorithm 1 (lines 1–6). Taking the variables as the vertices of a graph G, an edge is introduced between two vertices representing significantly interacted variables. We refer the interaction between two variables *statistically significant*, if the absolute value of the Pearson correlation coefficient (PCC) between them is greater than the critical value (denoted as σ in line 3) for PCC. The critical value depends on the p-value used and sample size. We compute the critical value in advance using p-value 0.05 and the given sample size. The function $pcc(u, v)$ in line 3 is used for calculating the absolute value of PCC between two variables u and v using data matrices $\mathbf{D_X}$ and $\mathbf{D_Y}$.

Identification of *CGRs* from the Graph. To identify *CGRs*, we first detect all maximal quasi-cliques from the graph. A γ-*quasi-clique* $Q = (V', E')$ is a subgraph of graph G, of which each vertex shares edges with at least $\lceil \gamma(d' - 1) \rceil$ other vertices in Q, where $d' = |V'|$ and $0 \leq \gamma \leq 1$ [12]. A *maximal* quasi-clique is a quasi-clique that is not contained inside any larger quasi-clique. We use the *Quick* algorithm [12] to find out all maximal quasi-cliques of G (line 7). *Quick* is a very fast and effective maximal quasi-cliques mining algorithm. We use different values for γ in different runs. The minimum size of the targeted quasi-cliques is set to 2θ, and the maximum size is set to m, where $m \geq 2\theta$ is a positive integer, chosen to get largest possible maximal quasi-cliques.

Each maximal quasi-clique Q_i is then split into two groups: qx_i and qy_i (line 10). Group qx_i is composed of the vertices representing the variables of $\mathbf{X}$, and similarly qy_i refers to the variables of $\mathbf{Y}$. We refer qx_i and qy_i as a *group pair*. CCA [8] is then used to get the strength of the collective relationship between a group pair (i.e. $\mathbf{r}$ in Definition 1). CCA is commonly used for quantifying the

---

Algorithm 1. *GRAPE* $(\mathbf{D_X}, \mathbf{D_Y}, k, \theta, \rho, \delta, \sigma)$

Input: $\mathbf{D_X}$ - an $(n \times p)$ matrix representing n samples of $\mathbf{X} = \{x_1, \ldots, x_p\}$,
   $\mathbf{D_Y}$ - an $(n \times q)$ matrix representing n samples of $\mathbf{Y} = \{y_1, \ldots, y_q\}$,
   k - top selection mark for *CGRs*, θ - minimum size threshold for *CGRs*,
   ρ - strength threshold, δ - membership similarity threshold, σ - significance level.
Output: $(C_1, \ldots, C_k)$ - top-k *CGRs*.
 1: Merge: $V \leftarrow \mathbf{X} \cup \mathbf{Y}$, $E \leftarrow \phi$
 2: **for all** $u, v \in V$ **do**
 3:    **if** $pcc(u, v) \geq \sigma$ **then**
 4:       $E \leftarrow E \cup (u, v)$
 5:    **end if**
 6: **end for**
 7: Call *Quick* on $G = (V, E)$ to get all maximal quasi-cliques $\{Q_1, \ldots, Q_{all}\}$, where
    $Q_i = \{(V_i, E_i) | V_i \subseteq V, E_i \subseteq E\}$
 8: $CGR_{all} \leftarrow \phi$
 9: **for** $i = 1$ to all **do**
10:    Spilt: $qx_i \leftarrow \{v \,|\, v \in V_i \wedge v \in \mathbf{X}\}$ and $qy_i \leftarrow \{v \,|\, v \in V_i \wedge v \in \mathbf{Y}\}$
11:    Compute $\mathbf{r}_i \leftarrow cca(qx_i, qy_i)$
12:    **if** $|qx_i| \geq \theta$, $|qy_i| \geq \theta$, and $\mathbf{r}_i \geq \rho$ **then**
13:       $CGR_{all} \leftarrow CGR_{all} \cup \{Q_i, \mathbf{r}_i\}$
14:    **end if**
15: **end for**
16: Sort $\{C_1, \ldots, C_t\} \in CGR_{all}$ in descending order of $\mathbf{r}_i$
17: $CGR_{rep} \leftarrow \phi$
18: **while** there is an unmarked *CGR* $C_i \in CGR_{all}$ **do**
19:    mark C_i
20:    $CGR_{rep} \leftarrow CGR_{rep} \cup C_i$
21:    **for each** unmarked *CGR* $C_j \neq C_i \in CGR_{all}$ **do**
22:       **if** $ms(V_i, V_j) \geq \delta$ **then**
23:          mark C_j
24:       **end if**
25:    **end for**
26: **end while**
27: **return** top-k *CGRs* $(C_1, \ldots, C_k) \in CGR_{rep}$

---

linear association between two sets of variables. Consider $\mathcal{A} = a'\mathbf{X}$, $\mathcal{B} = b'\mathbf{Y}$ be the corresponding linear combinations of sets of variables $\mathbf{X}$ and $\mathbf{Y}$ respectively, where a and b are coefficient vectors. Vectors a and b are chosen such that the correlation between $\mathcal{A}$ and $\mathcal{B}$, i.e.,

$$\mathbf{r} = Corr(\mathcal{A}, \mathcal{B}) = \frac{a' \Sigma_{XY} b}{\sqrt{a' \Sigma_{XX} a} \sqrt{b' \Sigma_{YY} b}} \tag{1}$$

is maximized, where Σ_{XX}, Σ_{YY} and Σ_{XY} are variance of $\mathbf{X}$, variance of $\mathbf{Y}$, and covariance between $\mathbf{X}$ and $\mathbf{Y}$, respectively. The correlation $\mathbf{r}$ between the pair of linear combinations in (1) is called *canonical correlation*. In line 11, the canonical correlations $\mathbf{r}_i$ between qx_i and qy_i are computed using the function $cca(qx_i, qy_i)$. Finally, only the (θ, ρ)-associated *CGRs* are stored (lines 12–14).

Selection of Top-k CGRs. We select representative *CGRs* based on membership similarity score and canonical correlation. We compute the membership similarity score of two sets of variables S and T using a modified version of Jaccard Index equation: $ms(S,T) = |S \cap T|/min\{|S|,|T|\}$ (line 22). The membership similarity score ranges between 0 and 1, where 1 denotes that S and T are the same, i.e. they contain the same members or variables. Now we define the representative *CGR* for two *CGRs* as follows:

Definition 3. *Given two CGRs* $(\mathcal{P}_i, \mathcal{E}_i, \mathbf{r}_i)$ *and* $(\mathcal{P}_j, \mathcal{E}_j, \mathbf{r}_j)$, *and membership similarity threshold* δ. *If* $ms(\mathcal{P}_i, \mathcal{P}_j) \geq \delta$ *and* $\mathbf{r}_i \geq \mathbf{r}_j$, *then* $(\mathcal{P}_i, \mathcal{E}_i, \mathbf{r}_i)$ *is referred to as the representative CGR.*

The representative *CGR* eliminates redundant *CGRs*, and thus it prevents number of *CGRs* to grow large. Using Definition 3, we select a set of representative *CGRs* (lines 16–26). From the representative *CGRs* set, we take the top-k *CGRs* based on canonical correlations, and return them as output (line 27).

3   Experimental Results

3.1   Experiments on Synthetic Datasets

We performed two simulation studies to investigate effectiveness of *GRAPE* in *CGRs* identification. Although we conducted several simulated runs using different sample sizes and different numbers of variables, due to page limit, in this paper we only report the more critical settings with small sample size.

In the first simulation study, we fixed sample size $n = 50$ and initially generated $p = 100$ variables for set $\mathbf{X}$ and $q = 150$ variables for set $\mathbf{Y}$ such that the first 10 variables of $\mathbf{X}$ and the first 10 variables of $\mathbf{Y}$ were correlated and the rest were independent noise. We then generated $\mathbf{X}$ and $\mathbf{Y}$ with gradually increased variable sizes up to $p = 4000$ and $q = 6000$. For each variable size setting, we performed 10 simulation runs.

The accuracy of identification was measured by *precision* and *recall*. Precision is the ratio of correctly identified variables of group pairs to the total identified variables for group pairs, while recall is the ratio of correctly identified variables of group pairs to the total variables in correlated group pairs. More specifically, for a cluster S, precision (denoted by P) and recall (denoted by R) are defined as: $P(S) = T_+(S)/(T_+(S) + F_+(S))$, and $R(S) = T_+(S)/(T_+(S) + F_-(S))$ respectively, where T_+, T_-, F_+, and F_- are true positives, true negatives, false positives, and false negatives in S, respectively.

We used this simulation to compare the performance of *GRAPE* with two sCCA methods, SCCA [15] and PMD [22]. For SCCA, we used 10-fold cross-validation to get optimal parameter values. In case of PMD, optimal parameters were obtained by using 10 permutations in each run. *GRAPE* achieved consistent higher precision than both SCCA and PMD, shown in Fig. 1 (Left). *GRAPE* suffered less in terms of recall compared to both SCCA and PMD for high dimensional data, given in Fig. 1 (Right). As both SCCA and PMD predicted more variables in case of high dimensional data, their recall suffered more.

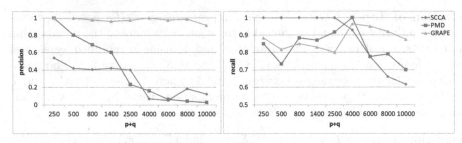

Fig. 1. Comparison of precision and recall

In the second simulation study, we fixed $n = 50$ and initialized $p = 100$ for $\mathbf{X}$ and $q = 150$ for $\mathbf{Y}$. The variables of $\mathbf{X}$ were divided into 25 groups of equal size, and similarly $\mathbf{Y}$ was split into 25 equal-sized groups. Then we created $\mathbf{X}$ and $\mathbf{Y}$ with higher number of variables up to $p = 4000$ and $q = 6000$, and the number of groups up to 500. For each variable size setting, all of the variables from $\mathbf{X}$ and $\mathbf{Y}$ were considered to be the vertices of a graph. For each sample, edges were placed independently at random between vertex pairs with probability more than 0.2 for an edge to fall between vertices in the same group and at best 0.1 to fall between vertices in different groups. We executed 10 simulation runs for each variable size setting.

In this study, we compared the effectiveness of canonical correlations as a criterion for ranking with that of modularity [14]. Assuming a graph G, the modularity can be expressed as a sum over the clusters as [7]:

$$\Omega = \sum_{c=1}^{n_c} \left[\frac{m_c}{m} - \left(\frac{d_c}{2m} \right)^2 \right] , \qquad (2)$$

where n_c is the number of clusters obtained by partitoning G, m_c is the number of edges joining vertices inside cluster c, and d_c is the sum of the degrees[1] of the vertices of c. Each summand in (2) stands for the contribution of c to the modularity of G. A cluster is 'good' if its contribution to modularity is positive. Higher positive contribution to modularity refers to better cluster, and higher positive modularity means better overall partition of G.

We used the number of correctly identified group pairs, and the *normalized mutual information (NMI)* [6] to compare the performance of ranking as illustrated in Fig. 2 (Left) and Fig. 2 (Right), respectively. In these column charts, it is evident that *GRAPE* constantly included more group pairs than the ranking by modularity, denoted as 'QuiM', meaning 'Quick+Modularity' (modularity metric was applied to the post-processed result of Quick algorithm to rank group pairs). Moreover, the NMI values of the group pairs identified by *GRAPE* were higher than the group pairs ranked by modularity.

---

[1] The *degree* of a vertex v is the number of edges incident to v.

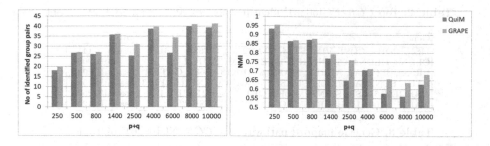

Fig. 2. Numbers of group pairs identified and NMI values for different variable sizes

3.2   Experiments on Real-World Datasets

We selected two real-world datasets: a biological NCI-60 dataset for Epithelial-to-Mesenchymal Transition (EMT), and a bibliography DBLP dataset.

NCI-60 Dataset for EMT. An EMT is a biological process that enables ep-ithelial cells to become migratory mesenchymal cells. The EMTs are associated with embryonic development, wound healing, organ fibrosis, and in the initiation of metastasis for cancer progression. The NCI-60 dataset includes microRNAs (miRNAs) expression profiles for the NCI-60 panel of 60 cancer cell lines [16][2]. miRNAs are known to play an essential role as regulators in post-transcriptional gene regulation. The messenger RNA (mRNA) expression profiles were down-loaded from ArrayExpress [3]. Cell lines categorized as epithelial (11 samples) and mesenchymal (36 samples) were used for this work. The differentially expressed gene analysis was performed using the limma package of Bioconductor [18], and 1635 mRNA probes and 43 miRNA probes were identified to be differentially expressed with p-value < 0.05 (adjusted p-value). Bootstrapping was used to overcome the problem of small number of samples.

The input for *GRAPE* was 47×1635 matrix *gene* and 47×43 matrix *mirna* of the 47 samples of miRNA and mRNA expression values, respectively. Ta-ble 2 shows the top *CGR* 'NCI60gp1' (with correlation 0.948) identified with 42 genes and 7 miRNAs for $\gamma = 0.85$. It is interesting to note that 'NCI60gp1' put members of the 'hsa-miR-200' family together.

We used GeneGo Metacore from GeneGo Inc. to identify the pathways pre-viously discovered in the literature that involve the genes in the identified top ranked *CGRs*. A *biological pathway* is a group of genes that participate in a particular biological process to perform certain functionality in a cell. To find controlling factors to a disease, it is more meaningful to study the genes by con-sidering their pathway information. Table 3 shows the first 12 pathways identified for 'NCI60gp1' of Table 2. It confirms that 'NCI60gp1' is highly relevant to the biological condition of the datasets. For instance, pathways number 1, 4, 8, 10

---

[2] Available at http://www.ncbi.nlm.nih.gov/geo/query/acc.cgi?acc=GSE26375

[3] http://www.ebi.ac.uk/arrayexpress, accession number E-GEOD-5720

Table 2. Top *CGR* 'NCI60gp1' identified for $\gamma = 0.85$ from NCI-60 dataset

Genes	miRNAs
MSN, CDH1, CDS1, VIM, ST14, STAP2, CCDC88A, CBLC	hsa-miR-141,
CLDN3, CLDN4, CLDN7, KRT19, EPCAM, F11R, LIX1L,	hsa-miR-200a,
PRSS8, SLC25A4, ANPEP, MAP7, MAPK13, CGN, GRTP1,	hsa-miR-200b,
LAD1, SCNN1A, LLGL2, CNKSR1, B3GNT3, ZNF165, TJP3,	hsa-miR-200c,
ARHGEF16, LSR, SPINT2, ELF3, QKI, C19orf21, RBM47,	hsa-miR-203,
RAB25, S100A14, BSPRY, MYO5C, MYO5B, ESRP2	hsa-miR-429

Table 3. GeneGo mapped pathways for *CGR* 'NCI60gp1' of Table 2

No	Pathway Maps	miRNAs and genes	*p*-value
1	Development_miRNA-dependent inhibition of EMT	hsa-miR-429, hsa-miR-200c, hsa-miR-200b, hsa-miR-200a, CDH1, hsa-miR-141, hsa-miR-200a-3p	1.423E-15
2	Cell adhesion_Tight junctions	TJP3, F11R, CGN, CLDN3, CLDN4, CLDN7	1.283E-09
3	Cell adhesion_Endothelial cell contacts by junctional mechanisms	F11R, CGN, VIM, CLDN3	1.320E-06
4	Development_TGF-β-dependent induction of EMT via MAPK	MAPK13, CDH1, VIM	4.659E-04
5	ENaC regulation in normal and CF airways	SCNN1A, PRSS8, ST14	6.642E-04
6	Cell cycle_Role of 14-3-3 proteins in cell cycle regulation	MAPK13, CDS1	2.303E-03
7	Cell adhesion_Gap junctions	TJP3, CGN	4.268E-03
8	Development_TGF-β-dependent induction of EMT via SMADs	CDH1, VIM	5.779E-03
9	Cytoskeleton remodelling_Keratin filaments	KRT19, VIM	6.106E-03
10	Development_TGF-β-dependent induction of EMT via RhoA, PI3K and ILK	CDH1, VIM	9.831E-03
11	Development_WNT signaling pathway. Part 2	CDH1, VIM	1.291E-02
12	Development_Regulation of EMT	CDH1, VIM	1.847E-02

and 12 are direct pathways of the development of EMT, and others are important pathways involved in the process of EMT. Moreover, pathway number 1
includes total 12 members, of which 7 were identified in 'NCI60gp1'.

DBLP Dataset. The DBLP dataset [19] [4] consists of 1783 papers published
in 7 conferences from 1995 to 2004. The information extracted from the papers
were organized into set *Author* containing 3036 authors and set *Terms* having
2090 terms (words were extracted from the titles based on tf-idf, and then words
with low df were removed). Table 4 depicts the top 5 *CGRs* identified from the
DBLP dataset for $\gamma = 0.4$. The 1st *CGR*, gp1 covered 3 papers associated with
patterns. Again, the 5th *CGR*, gp5 fetched into terms from 2 different papers,
where the authors are co-authors of *B. Zhang* and *W. Fan*.

It is interesting to note that terms and authors were *uniformly* included in
a *CGR* based on collective interactions. For example, the 9th *CGR*, gp9 (with
correlation 0.822) in Fig. 3 combined 10 authors and 16 terms from 5 papers.
These papers were published with the contribution of total 11 co-authors, only
T. Mogawa was missed out. The reason behind the exclusion is less association
of *T. Mogawa* in the collective interactions of authors and terms in gp9.

---

[4] Downloaded from `http://leitang.net/data/dblp.tar.gz` in October 2013.

Table 4. Top 5 *CGRs* identified for $\gamma = 0.4$ from DBLP dataset

No	CC	#A	#T	Authors	Terms
gp1	0.9999067	12	12	T. Miyahara, D. Nauck, ...	tag, patterns, ...
gp2	0.9991713	5	9	D. Nauck, T.P. Martin, ...	association, care, ...
gp3	0.9834444	14	11	P. Blair, K. Sarkar, ...	distributed, automated, ...
gp4	0.9670259	14	10	T.W. Finin, Y. Peng, ...	services, semantic, ...
gp5	0.9398762	16	8	W. Fan, B. Zhang, ...	citation, criterion, ...

Author (10):
K. Furukawa, T. Shoudai, K. Yamada, S. Hirokawa, Y. Nakamura, K. Takahashi, T. Miyahara, H. Ueda, T. Uchida, Y. Suzuki

Terms (16):
characteristic, contractible, discovery, documents, extracting, frequent, irregular, maximally, patterns, semistructured, structured, structures, tag, tree, variables, words

Papers (5):
T. Miyahara, T. Shoudai, T. Uchida, K. Takahashi, H. Ueda
Polynomial Time Matching Algorithms for **Tree**-Like **Structured Patterns** in Knowledge **Discovery.**
PAKDD 2000;

K. Furukawa, T. Uchida, K. Yamada, T. Miyahara, T. Shoudai, Y. Nakamura
Extracting Characteristic Structures among **Words** in **Semistructured Documents.**
PAKDD 2002;

T. Uchida, T. Mogawa, **Y. Nakamura**
Finding **Frequent** Structural Features among **Words** in **Tree-Structured Documents**.
PAKDD 2002;

T. Miyahara, Y. Suzuki, T. Shoudai, T. Uchida, S. Hirokawa, K. Takahashi, H. Ueda
Extraction of **Tag Tree Patterns** with **Contractible** Variables from **Irregular Semistructured** Data.
PAKDD 2003;

T. Miyahara, Y. Suzuki, T. Shoudai, T. Uchida, K. Takahashi, H. Ueda
Discovery of **Maximally Frequent Tag Tree Patterns** with **Contractible** Variables from **Semistructured Documents**.
PAKDD 2004;

Fig. 3. Details of one of the top-10 *CGRs* identified from DBLP datasets by *GRAPE*

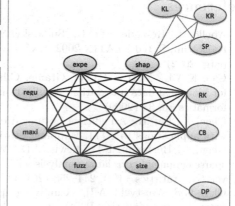

Authors (6):
Christian Borgelt, Rudolf Kruse, S. Prabhakar, Karthik Ramani, Kuiyang Lou, David Poole.

Words (6):
regularization, maximization, expectation, size, shape, fuzzy.

Papers (3):
Christian Borgelt, Rudolf Kruse
Shape and **Size Regularization** in **Expectation Maximization** and **Fuzzy** Clustering.
PKDD 2004;

Kuiyang Lou, S. Prabhakar, Karthik Ramani
Content-based Three-dimensional Engineering **Shape** Search.
ICDE 2004;

David Poole
Estimating the **size** of the telephone universe: a Bayesian Mark-recapture approach.
KDD 2004;

Fig. 4. Details of one of the top-10 groups identified from DBLP datasets by [13]

We compared our result with that of [13], in which a smaller subset of this dataset was used. The dataset had 2022 authors and 1650 terms extracted from 1071 papers published in 5 conferences from 2000 to 2004. The method identified 143 groups, many of which showed *skewed* interactions. For eaxmple, all terms of 4th group (with correlation 0.994) given in Fig. 4 (Left) were extracted from the 1st paper alone with a term each found common in other papers. The links in Fig. 4 (Right) illustrates *skewness* instead of *uniform smoothness* in interactions among nodes (the first 4 letters for terms and abbreviated names for authors were used in labelling nodes). This skewness may sometimes combine unrelated terms and authors together. Moreover, the method only allowed disjoint groups, whereas *GRAPE* handled overlapping groups.

4   Conclusion

In this paper, we have defined the notation of collective group relationships (*CGRs*), and have proposed a method to discover *CGRs* from heterogeneous datasets. The central idea of our proposed method is to integrate CCA with quasi-clique mining in order to enrich the identification of groups with both structural information and strength of relationships. We have experimented on two simulation settings as well as real-world biological NCI-60 dataset and bibliography DBLP dataset. The accuracy of the method is comparatively high, and it has the ability to identify overalpping groups. The experimental results have demonastrated that the proposed method is able to reveal correct group information with links and the strength of collective relationships, and provide useful insights into the structure and functionality of the systems.

Acknowledgments. This work was partially supported by Australian Research Council Discovery grant DP130104090.

References

1. Abello, J., Resende, M.G.C., Sudarsky, S.: Massive Quasi-Clique Detection. In: Rajsbaum, S. (ed.) LATIN 2002. LNCS, vol. 2286, pp. 598–612. Springer, Heidelberg (2002)
2. Cao, K.A.L., Martin, P.G.P., Granié, C.R., Besse, P.: Sparse canonical methods for biological data integration: Application to a cross-platform study. BMC Bioinformatics 10, 34 (2009)
3. Chen, X., Liu, H.: An efficient optimization algorithm for structured sparse CCA, with applications to eQTL Mapping. Statistics in Biosciences 4(1), 3–26 (2012)
4. Chen, J., Bushman, F.D., Lewis, J.D., Wu, G.D., Li, H.: Structure-constrained sparse canonical correlation analysis with an application to microbiome data analysis. Biostatistics 14(2), 244–258 (2013)
5. Chiu, G.S., Westveld, A.H.: A unifying approach for food webs, phylogeny, social networks, and statistics. PNAS 108(38), 15881–15886 (2011)
6. Danon, L., Díaz-Guilera, A., Duch, J., Arenas, A.: Comparing community structure identification. Journal of Statistical Mechanics: Theory and Experiment P09008 (2005)

7. Fortunato, S.: Community detection in graphs. Physics Reports 486, 75–174 (2010)
8. Hotelling, H.: Relations Between Two Sets of Variates. Biometrika 28(3/4), 321–377 (1936)
9. Jain, A.K.: Data clustering: 50 years beyond K-means. Pattern Recognition Letters 31(8), 651–666 (2010)
10. Lee, W., Lee, D., Lee, Y., Pawitan, Y.: Sparse Canonical Covariance Analysis for High-throughput Data. Statistical Applications in Genetics and Molecular Biology 10(1): Article 30 (2011)
11. Lin, D., Zhang, J., Li, J., Calhoun, V.D., Deng, H.W., Wang, Y.P.: Group sparse canonical correlation analysis for genomic data integration. BMC Bioinformatics 14, 245 (2013)
12. Liu, G., Wong, L.: Effective Pruning Techniques for Mining Quasi-Cliques. In: Daelemans, W., Goethals, B., Morik, K. (eds.) ECML PKDD 2008, Part II. LNCS (LNAI), vol. 5212, pp. 33–49. Springer, Heidelberg (2008)
13. Liu, H., Li, J., Liu, L., Liu, J., Lee, I., Zhao, J.: Exploring Groups from Heterogeneous Data via Sparse Learning. In: Pei, J., Tseng, V.S., Cao, L., Motoda, H., Xu, G. (eds.) PAKDD 2013, Part I. LNCS (LNAI), vol. 7818, pp. 556–567. Springer, Heidelberg (2013)
14. Newman, M.E.J., Girvan, M.: Finding and evaluating community structure in networks. Physical Review E 69(2), 26113 (2004)
15. Parkhomenko, E., Tritchler, D., Beyene, J.: Sparse Canonical Correlation Analysis with Application to Genomic Data Integration. Statistical Applications in Genetics and Molecular Biology, 8(1), Article 1 (2009)
16. Søkilde, R., Kaczkowski, B., Podolska, A., Cirera, S., Gorodkin, J., Møller, S., Litman, T.: Global microRNA Analysis of the NCI-60 Cancer Cell Panel. Molecular Cancer Therapeutics 10, 375–384 (2011)
17. Soneson, C., Lilljebjörn, H., Fioretos, T., Fontes, M.: Integrative analysis of gene expression and copy number alterations using canonical correlation analysis. BMC Bioinformatics 11, 191 (2010)
18. Smyth, G.K.: Limma: linear models for microarray data. Statistics for Biology and Health. Bioinformatics and Computational Biology Solutions using R and Bioconductor. pp. 397-420. Springer (2005)
19. Tang, L., Liu, H., Zhang, J., Nazeri, Z.: Community Evolution in Dynamic Multi-Mode Networks. In: 14th ACM SIGKDD Conference on Knowledge Discovery and Data Mining (KDD), Las Vegas, USA, pp. 677–685 (2008)
20. Waaijenborg, S., Zwinderman, A.H.: Sparse canonical correlation analysis for identifying, connecting and completing gene-expression networks. BMC Bioinformatics 10, 315 (2009)
21. Wagner, G.P., Pavlicev, M., Cheverud, J.M.: The Road to Modularity. Nature Reviews Genetics 8(12), 921–931 (2007)
22. Witten, D., Tibshirani, R., Hastie, T.: A Penalized Matrix Decomposition, with Applications to Sparse Principal Components and Canonical Correlation Analysis. Biostatistics 10(3), 515–534 (2009)
23. Yan, J.J., Zheng, W., Zhou, X., Zhao, Z.: Sparse 2-D canonical correlation analysis via low rank matrix approximation for feature extraction. IEEE Signal Process Letters 19(1), 51–54 (2012)
24. Yeung, K.Y., Medvedovic, M., Bumgarner, R.E.: From co-expression to co-regulation: how many microarray experiments do we need? Genome Biology, 5(7), Article R48 (2004)

Efficiently Retrieving Top-k Trajectories by Locations via Traveling Time

Yuxing Han[1], Lijun Chang[2], Wenjie Zhang[2], Xuemin Lin[1,2], and Liping Wang[1]

[1] East China Normal University, China
[2] The University of New South Wales, Australia
sei.yxhan@gmail.com, {ljchang,zhangw,lxue}@cse.unsw.edu.au,
lipingwang@sei.ecnu.edu.cn

Abstract. The flourishing industry of location-based services has collected a massive amount of users' positions in the form of spatial trajectories, which raise many research problems. In this paper, we study a trajectory retrieving query, *k-TLT*, which aims at retrieving the top-*k Trajectories* by *Locations* and ranked by *traveling Time*. Given a set Q of query locations, a *k-TLT* query retrieves top-*k* trajectories that are close to Q with respect to traveling time. In contrast to existing works which consider only location information, *k-TLT* queries also consider the traveling time information, which have many applications, such as travel route planning and moving object study. To efficiently answer a *k-TLT* query, we first online compute a list L_q of trajectories for each query location $q \in Q$, such that trajectories in L_q are ranked by their traveling time to q. Based on the online generated lists L_q corresponding to query locations, a small set of candidate trajectories that are close to Q is selected by iteratively retrieving trajectories from lists L_q. Then, the set of candidate trajectories is refined and pruned to determine the top-*k* trajectories. We conduct extensive experiments on a real trajectory dataset and verify the efficiency of our approach.

Keywords: Trajectory retrieving, locations, traveling time, efficiency.

1 Introduction

As a result of an abundant number of trajectory data generated by mobile devices and global position systems, various innovative location-based applications have been studied. Among them, one prominent application is retrieving interesting trajectories. The existing works on trajectory retrieving either query by a trajectory [2,3,10] or by locations [4,12]. The works on retrieving similar trajectories by a query trajectory focus on defining similarities between trajectories [2,3,10]. We study querying trajectories by locations in this paper. For querying trajectories by locations, the existing works [4,12] compute trajectories that are geographically close to a set of query locations while neglecting road conditions and traveling time information. However, in many real applications, the set of geographically close trajectories may not be the best choice due to rough road conditions, and we should also regard traveling time information as an important factor. Imagine a person who takes his first trip to an unfamiliar city but with a limited travel time for some personal issues. In order to save time, he chooses to travel between attractive places by taxi, and hopes to spend less time on traveling and have

H. Wang and M.A. Sharaf (Eds.): ADC 2014, LNCS 8506, pp. 122–134, 2014.
© Springer International Publishing Switzerland 2014

more time enjoying himself in the places that he really likes. Here, the attractive places that he wants to visit can be regarded as query locations, and candidate traveling routes are existing trajectories. Historical trajectories recorded in the database could surely provide similar routes that help him to plan this trip. This motivates us to study a new trajectory retrieving query, *k-TLT*, which aims at retrieving the top-*k* trajectories by locations and ranked by traveling time. Given a set Q of query locations, a *k-TLT* query retrieves the top-*k* trajectories that are close to Q with respect to traveling time.

There are two challenges in efficiently answering a *k-TLT* query: 1) the number of trajectories and trajectory points is very large such that processing all trajectories is time-consuming; 2) the existing index structure for trajectory points, R-tree [7], considers only distance information. To tackle these challenges, we first study how to incorporate speed information of trajectory points into an R-tree, such that, given a query location q, trajectory points can be efficiently retrieved in non-decreasing order with respect to their traveling time to q. The traveling time (*reach time*) between a trajectory and a location q (resp. a set Q of locations) is the minimum traveling time from any trajectory point to q (resp. the sum of traveling time to locations in Q). Based on the augmented R-tree, we online compute a list L_q of trajectories for each query location q such that trajectories in L_q are ranked by their traveling time to q. Based on the online generated lists L_q, one corresponding to each query location $q \in Q$, we propose techniques to select a small set of candidate trajectories that are close to Q by iteratively retrieving trajectories from lists L_q. Here, we use the *bitsets* technique to effectively record statistics of trajectories. Then, we iteratively refine and prune trajectories in the candidate set to get the top-*k* trajectories.

The rest of this paper is organized as follows. Section 2 gives our problem statement, and we present our new efficient algorithm for answering *k-TLT* queries in Section 3. Experimental results are reported in Section 4, followed by related work in Section 5. We conclude the paper in Section 6.

2 Problem Definition

A *spatial trajectory R* is a trace generated by a moving object in geographical spaces, and usually represented by a time-ordered sequence of location points, (t_1, p_1), (t_2, p_2), $\ldots, (t_N, p_N)$, with $t_1 < t_2 < \cdots < t_N$. Here, N is the number of recorded points in R, and each location point p_i is represented by (*longitude*, *latitude*) corresponding to time t_i. In order to compute the traveling time between a trajectory and a location point, we precompute an average speed at each location point p_i in R, denoted as s_i. Therefore, in the following, we consider a trajectory R to be represented by $(t_1, p_1, s_1), (t_2, p_2, s_2)$, $\ldots, (t_N, p_N, s_N)$, where s_i is the speed at location p_i, and (t_i, p_i, s_i) is referred to as a *trajectory point e_i of R*.

Definition 1. *The **reach time** between a trajectory point e and a location point q, denoted single_time(e, q), is the traveling time from e.p to q, i.e.,*

$$single\_time(e, q) = \frac{d(e.p, q)}{e.s},$$

where d(,) is the Euclidean distance between two location points.

Definition 2. *The **reach time** between a trajectory R and a location point q, denoted t(R, q), is the minimum reach time between any point in R and q, i.e.,*

$$t(R, q) = \min_{e \in R} single\_time(e, q)$$

*Here, we say that q is **matched** to $e_i.p$, where e_i has the minimum reach time to q among all trajectory points in R.*

Definition 3. *Given a set Q of location points $\{q_1, q_2, \ldots, q_m\}$, the **reach time** between R and Q, denoted T(R, Q), is the sum of reach time between R and each point in Q, i.e.,*

$$T(R, Q) = \sum_{q \in Q} t(R, q)$$

Fig. 1 illustrates an example of computing reach time between a trajectory R and a set Q of three points $\{q_1, q_2, q_3\}$. q_1 and q_3 are respectively matched to p_2 and p_5, which are their geographically closest points in R. q_2 is matched to p_4 due to less traveling time from p_4 to p_2. Although p_3 is geographically closer to q_2 than p_4, it may take more time to travel from p_3 to p_2 due to rough road conditions around p_3. Therefore, the reach time between R and Q is $T(R, Q) = t(R, q_1) + t(R, q_2) + t(R, q_3) = 7min$.

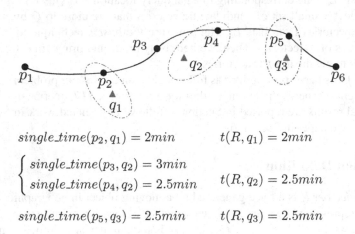

$single\_time(p_2, q_1) = 2min$ $t(R, q_1) = 2min$

$\begin{cases} single\_time(p_3, q_2) = 3min \\ single\_time(p_4, q_2) = 2.5min \end{cases}$ $t(R, q_2) = 2.5min$

$single\_time(p_5, q_3) = 2.5min$ $t(R, q_3) = 2.5min$

Fig. 1. Reach Time Computation

Problem Statement. Given a trajectory database D and a query Q consisting of m location points, $Q = \{q_1, \ldots, q_m\}$, we study the problem of *k-TLT* query, which aims at retrieving the top-**k** Trajectories by Locations and ranked by traveling Time. That is, a *k-TLT* query retrieves from D a subset K consisting of k trajectories whose *reach time* to Q is not more than any trajectories in (D−K), i.e., $\max_{R_i \in K} T(R_i, Q) \leq \min_{R_j \in D-K} T(R_j, Q)$.

Table 1 lists the notations and symbols used in this paper.

3   Query Processing

Naive Approach. For a *k-TLT* query Q, where trajectory points of trajectories in D are stored in an R-tree [7], a naive approach would be computing the reach time between

Table 1. List of Notations

Notations	Explanation
R, D	a trajectory, and a trajectory database
e	a trajectory point
Q, m	the set (resp. number) of query locations
$d(,)$	Euclidean distance between two location points
$t(,)$	reach time between a trajectory and a query location
$T(,)$	reach time between a trajectory and a query set of locations
k, K	the number k, and the top-k trajectories
C	a candidate set of trajectories

Algorithm 1. Naive Approach

Input: a trajectory database D, a set of query locations Q, and k
Output: the top-k trajectories in D with respect to reach time to Q
1 **foreach** *trajectory R in D* **do**
2     $\lfloor$ compute the reach time $T(R, Q)$ between R and Q;

3 sort all trajectories in D according to their reach time to Q;
4 **return** the top-k trajectories;

---

every trajectory R in D and Q, then sorting trajectories by their reach time to Q, and returning the top-k trajectories, as shown in Algorithm 1.

The naive approach is time-consuming and impractical. This is because it needs to compute the reach time between every trajectory in D and Q, although we are only interested in the top-k trajectories. Usually, k is much smaller than the number of trajectories in D and is furthermore much smaller than the number of trajectory points in D. Therefore, we propose an incremental approach to computing top-k trajectories by retrieving only a small set of trajectory points and then a small set of trajectories from D in the following.

Incremental Approach. The general idea of our incremental approach is as follows. 1) For each query location q in Q, we retrieve trajectory points from D in non-decreasing order with respect to their reach time to q; therefore, trajectories are retrieved from D in non-decreasing order with respect to their reach time to q, and we consider the sorted trajectories according to q as a sorted list L_q. 2) Based on the online generated lists L_q, we select a small set of candidate trajectories which contains the top-k trajectories. 3) Finally, we refine and prune trajectories in the candidate set to get the top-k trajectories. The pseudocode is shown in Algorithm 2.

Based on the incremental approach, we compute top-k trajectories by visiting only a small set of trajectory points and trajectories from D, which is much more efficient than the naive approach in Algorithm 1. We discuss the three steps of our incremental approach in the following three subsections, respectively.

Algorithm 2. Incremental Approach

Input: a trajectory database D, a set of query locations Q, and k
Output: the top-k trajectories in D with respect to reach time to Q
1 **foreach** *query location q in Q* **do**
2    $\lfloor$ initialize and prepare a sorted list L_q for q;

3 compute a set C of candidate trajectories based on L_q for all $q \in Q$;
4 refine and prune trajectories in C to get the top-k trajectories K;
5 **return** K;

3.1 Online Generating List L_q

Augmenting R-tree. In order to retrieve trajectory points from D in non-decreasing order with respect to their reach time to a query location q, we need to organize trajectory points in D in a data structure, and R-tree [7] is a good choice for retrieving points in sorted order. However, R-tree considers only distance information, while our purpose is to rank points with respect to their traveling time which considers both the distance and the speed information. Therefore, we propose to also incorporate the speed information into the R-tree data structure for fast query processing.

The key idea of R-tree [7] is to group nearby objects and represent them with their *minimum bounding rectangle (MBR)* in the next higher level of the tree. In order to retrieve trajectory points in non-decreasing order with respect to their traveling time to a query location, we incorporate the speed information into the R-tree, such that each rectangle is also associated with a speed information. For a rectangle at the leaf level, the speed information is the maximum speed of the contained trajectory points; while at higher levels the maximum speed stored in its children rectangles, or equivalently the maximum speed of trajectory points contained by the rectangle. The construction of the augmented R-tree $\mathbb{R}$ is the same as that in [7] by assigning the speed information in a bottom-up fashion as post-processing.

Definition 4. *Given a query location q, a trajectory point e in trajectory R is a **match point** if q is matched to e in the reach time between R and q, i.e., e has the minimum reach time to q among all trajectory points in R.*

Online Generating List L_q. Based on the augmented R-tree $\mathbb{R}$, for a query location q, we can retrieve trajectory points from $\mathbb{R}$ incrementally with respect to their reach time to q, and add the corresponding trajectory to L_q if the trajectory point is a *match* point. Recall that, L_q consists of trajectories and trajectories in L_q are sorted in non-decreasing order with respect to their reach time to q.

The algorithm to online generate list L_q is shown in Algorithm 3, which conducts a *best first search* on the augmented R-tree $\mathbb{R}$. We maintain a *priority queue* $\mathbb{H}_q$ to store the rectangles and trajectory points visited during the search process. For a trajectory point in $\mathbb{H}_q$, its key is the reach time between the trajectory point and q; while for a rectangle, its key is a lower bound of the reach time between any trajectory points contained in the rectangle and q. The lower bound of reach time is computed as *min_dist/speed*, where *min_dist* is the minimum distance between the rectangle and q, and *speed* is the speed information stored at the rectangle. Initially, H_q contains the root node of $\mathbb{R}$. Then, entries r are iteratively popped from H_q (Line 4). If r is a location point, we add the trajectory corresponding to r to L_q (Lines 5-7). Otherwise, we push to $\mathbb{H}_q$ either the children rectangles of r (Lines 8-10) or the location points contained in r (Lines 11-13), depending on whether r is a leaf node or not.

Algorithm 3. Online Generating L_q

   Input: a query location q, an R-tree $\mathbb{R}$.
1 $\mathbb{H}_q \leftarrow \emptyset$;
2 Enheap($\mathbb{H}_q$, $\mathbb{R}.root\_node$, 0);
3 **while** $\mathbb{H}_q$ *is not empty* **do**
4      $r \leftarrow$ Deheap($\mathbb{H}_q$);
5   **if** *r is a location point* **then**
6   **if** *r is match point of q* **then**
7   Add the trajectory corresponding to r to L_q;
8   **else if** *r is a non-leaf node* **then**
9   **foreach** *child node c of r in* $\mathbb{R}$ **do**
10   Enheap($\mathbb{H}_q$, c, $c.min\_dist/c.speed$);
11   **else**
12   **foreach** *trajectory point e in r* **do**
13   Enheap($\mathbb{H}_q$, e, $d(e.p, q)/e.s$);

Note that, in our approach to answering a k-*TLT* query, we do not generate the entire list L_q, but one trajectory only. That is, we construct L_q on-demand, and each time only one trajectory is computed and added to L_q. Once L_q becomes empty, we compute the next trajectory.

3.2 Generating Candidate Set C

Based on the online generated lists L_q, the main idea of generating candidate set C is iteratively retrieving trajectories R from L_q and adding R to C, and terminating once there are at least k *all-covering* trajectories in C. A trajectory R is called an *all-covering* trajectory if it has been retrieved from all lists L_q, $\forall q \in Q$; that is, the reach time between R and Q is already computed. To achieve this, we construct a fixed-sized *priority queue* $\mathbb{Q}$, which contains m entries, one corresponding to each query location q, i.e., $(q, top(L_q))$. $top(L_q)$ denotes the top trajectory in L_q, and the key of $(q, top(L_q))$ in $\mathbb{Q}$ is the reach time between $top(L_q)$ and q which is stored in L_q.

The pseudocode is shown in Algorithm 4. To construct the candidate set C, we iteratively pop the top entry (q, R) from $\mathbb{Q}$ (Line 4), i.e., R is the current top trajectory in L_q. Then R is added to C and removed from L_q, and the next top trajectory in L_q is pushed into $\mathbb{Q}$ (Lines 5-6). The algorithm terminates once the *AllCoverTest* of C returns true (Lines 7-8), i.e., there are at least k all-covering trajectories in C. The correctness of Algorithm 4 directly follows from the following lemma.

Algorithm 4. Generating Candidate Set C

Input: Query Q, k, and m online generated lists L_q
Output: Candidate Set C

1   $C \leftarrow \emptyset$;
2   initialize a priority queue $\mathbb{Q}$ to contain $(q, top(L_q))$ for each $q \in Q$;
3   **while** *true* **do**
4   Pop an entry (q, R) from $\mathbb{Q}$;
5   Remove R from L_q, and push $(q, top(L_q))$ to $\mathbb{Q}$;
6      $C \leftarrow C \cup \{R\}$;
7   **if** *AllCoverTest(C) is true* **then**
8   **return** C;

Lemma 1. *Given a k-TLT query Q, the set C of trajectories returned by Algorithm 4 contains the top-k trajectories in D that are closest to Q with respect to their reach time to Q.*

Proof. We prove by contradiction. Assume that there is such a trajectory R in the top-k trajectories for a query Q that is not in C. There are at least k all-covering trajectories in C, let R' be the one with minimum reach time to Q. Since $T(R, Q) < T(R', Q)$, there must exist once query location q such that $T(R, q) < T(R', q)$. Then R should be retrieved from L_q before R', which means that R is in C. Contradiction. Thus, the lemma holds.   □

Implementing AllCoverTest. In Algorithm 4, we need to conduct all cover test by invoking procedure *AllCoverTest*. One naive approach is checking all trajectories in C to count the number of all-covering trajectories. However, this is time consuming, since we need to conduct the test each time when we retrieved a trajectory from any list L_q.

To efficiently count the number of *all-covering* trajectories, we propose a bitset-based implementation of *AllCoverTest*. For each trajectory in C, we construct a bitset with m bits, one corresponding to each query location, which are initialized to be 0. Note that, here, we do not construct bitsets for trajectories not in C. We use a hash table to store trajectories in C. The number of query locations is usually small in practice, then the size of all bitsets constructed is very small. Whenever a trajectory R is added to C (Line 6), the corresponding bit of the bitset of R is set to 1. A trajectory becomes all-covering, if all m bits are set to be 1; that is, by regarding the bitset as a binary number, the bitset has value $(2^m - 1)$ which can be tested in constant time. In Algorithm 4, we incrementally maintain the number of all-covering trajectories in C, which can be maintained in constant time by checking whether the newly added trajectory R at Line 6 is an all-covering trajectory. Then, Line 7 runs in constant time.

In addition to fast count the number of all-covering trajectories, the bitset technique can also be used in checking match point at Line 6 of Algorithm 3. For example, we first check whether the trajectory R corresponding to r is in C, if yes, then the bit corresponding to query location q of the bitset of R indicates whether r is a match of q or not; otherwise, r is a match of q. Therefore, Line 6 of Algorithm 3 also runs in constant time. The bitset technique adopted here actually greatly improves the performance of our algorithm, as shown in our experimental study (Sec 4.1).

3.3 Refining and Verifying C

A naive approach to verifying the candidate set C is computing the reach time between every trajectory in C and Q, and then returning the top-k trajectories. However, this is still time-consuming, especially for trajectories containing a lot of trajectory points. Now, let's consider the trajectories in C that are not *all-covering* ones. In fact, we have already get the information of reach time between such trajectories R and certain query locations q, such that R is retrieved from L_q. Therefore, we can use this information to compute a lower bound of the reach time between R and Q, and then prune the trajectory R if its lower bound reach time is not smaller than the largest reach time of the current top-k candidate trajectories. The lower bound reach time between a trajectory R and Q is computed following from the lemma below.

Lemma 2. *For a trajectory R in C that is not an all-covering trajectory, the low bound of its reach time to query Q is computed as:*

$$LB(R, Q) = \sum_{q \in Q_c} t(R, q) + \sum_{q' \in Q - Q_c} t(R_{q'}, q'),$$

where Q_c is the subset of query locations that R has been covered in the candidate set C, and $R_{q'}$ is the last retrieved trajectory from list $L_{q'}$.

Proof. The lemma directly follows from the definition of reach time between R and Q, since $t(R, q') \geq t(R_{q'}, q')$ for each $q' \in Q - Q_c$.   □

Algorithm 5. Refine & Verification

Input: Query locations Q, Candiate Set C, and priority queue $\mathbb{Q}$
Output: K: top-k trajectories

1 $K \leftarrow$ the top-k *all-covering* trajectories in C;
2 $\theta \leftarrow$ GetLargest(K);
3 **foreach** *non-all-covering trajectory R in C* **do**
4   compute $LB(R, Q)$;
5   **if** $LB(R, Q) < \theta$ **then**
6   compute $T(R, Q)$;
7   **if** $T(R, Q) < \theta$ **then**
8   replace the trajectory with the largest *reach time* in K by R;
9              $\theta \leftarrow$ GetLargest(K);

10 **return** K;

Based on Lemma 2, the algorithm to refine and verify trajectories in C is shown in Algorithm 5. The result set K is first initialized to contain the top-k all-covering trajectories in C (Line 1), and the largest reach time θ for trajectories in K is obtained by the *GetLargest* procedure (Line 2). Then, we iteratively verify each non-all-covering trajectory in C (Line 3-9). For each non-all-covering trajectory R, we first compute its lower bound reach time to Q, and R is pruned if $LB(R, Q) \geq \theta$; otherwise, the reach time between R and Q is computed, and K is updated by R if $T(R, Q) < \theta$. After processing all non-all-covering trajectories in C, the k trajectories in K are the top-k trajectories for the *k-TLT* query Q.

4   Experimental Evaluation

We conduct experiments on a real taxi dataset generated by Microsoft Research Asia (MSRA)[14,15] to verify the efficiency of our proposed algorithms. This dataset contains trajectories generated by $33,000$ taxis over 3 months within Beijing. There are total $69,541$ trajectories and $1,644,009$ trajectory points. The query locations are generated manually by selecting a sequence of coordinates of places of interests that complies with a reasonable visiting order. We do not use random generation of query locations, because random generation will cause a sudden jump from one location to another location which rarely happens in real life. In our experiments, we vary k, the number of returned trajectories, and $|Q|$, the number of query locations, and report the running time of algorithms. All the experiments are performed on a machine with an Intel i5 CPU(3.10GHz) and 8GB main memory, running Windows 7.

4.1   Evaluating Total Running Time

We evaluate the performance of the algorithms via two metrics — *Execution Time* and *Node Access* of R-tree. *Execution Time* is the total running time of algorithms, and *Node Access* is the number of R-tree nodes accessed in the algorithms. We compare two approaches: the naive approach, denoted *Naive*, and the incremental approach, as discussed in Section 3. For the incremental approach, we compared two implementations of the *AllCoverTest* in Algorithm 4: the naive way by checking all trajectories, denoted *Incremental*, and the bitset-based implementation, denoted *Incremental-Bitset*.

Fig. 2 shows experimental results on *Execution Time* for the three algorithms. Overall, the two incremental algorithms are much faster than the naive algorithm under different values of k and $|Q|$. When varying k, Fig. 2(a) illustrates that the bitset-based implementation significantly improves the performance by one order of magnitude towards all k values. When varying $|Q|$, obviously, the bitset-based implementation also significantly outperforms the naive all cover test method as shown in Fig. 2(b). As shown in Fig. 2(b), when $|Q|$ becomes larger, the running time of the naive *AllCoverTest* method increases almost exponentially. This is because that the search space grows very fast as the number of query points increases. However, bitset-based implementation doesn't need to check every trajectory for an *AllCoverTest* by means of bitwise operations, which shows its advantage.

Experimental results on *Node Access* are given in Fig 3. Because different implementations of *AllCoverTest* do not affect the number of R-tree nodes accessed, we only compare Node Access between *Naive* and *Incremental-Bitset*. When query locations

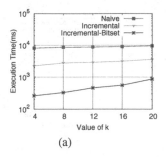
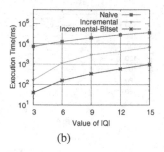

Fig. 2. Execution Time

Q is chosen, *Node Access* of *Naive* fixes at a very large number as shown in Fig 3(a). When $|Q|$ becomes larger, *Node Access* of *Naive* grows very quickly(Fig 3(b)). This is because *Naive* needs to access all R-tree nodes for each query location to compute the reach time between trajectories and a query location. As expected, our proposed incremental approach accesses only a small subset of R-tree as shown in Fig 3, which is benefited from our incremental manner to online generate the ranked lists L_q for each query location q.

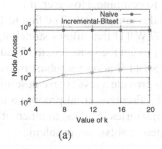
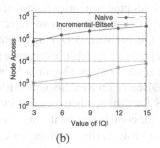

Fig. 3. Node Access

4.2 Evaluating Refine and Verification

We compare the performances of different approaches in the refine and verification stage of our algorithm. More specifically, we compare two approaches, *No Pruning* which computes a reach time for each trajectory in the candidate set C, and *Prune with LB* as discussed in Section 3.3. We show the running time in Fig. 4. As expected, the strategy of pruning with LB runs faster than the one without pruning. Note that we have already pruned an abundant number of un-qualified trajectories at the candidate generation stage. At the refinement stage, we only need to verify trajectories in a relatively small candidate set C. Nevertheless, the pruning strategy effectively prunes a lot of candidate trajectories in C without actually computing their reach time to Q; therefore, results in much smaller running time, as shown in Fig. 4.

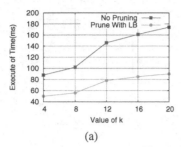 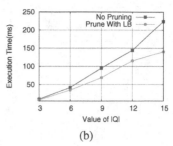

(a)　　　　　　　　　　　　　　　　　　(b)

Fig. 4. Refine & Verification

5 Related Work

Many algorithms towards searching points or regions have been proposed [8,9,11], however, they cannot be applied to retrieving trajectories. The existing works towards retrieving trajectories can be classified into two categories: 1) trajectory retrieving by a query trajectory; 2) trajectory retrieving by locations.

Trajectory Retrieving by a Trajectory. Here, the query is a trajectory, which finds similar trajectories in a database to the query trajectory. To tackle the problem, several similarity functions have been proposed in the literature. Euclidean distance is a commonly adopted similarity function, and it is easy to compute. However, it has the disadvantage that requires the two trajectories to have the same length. The first technique to solve this problem is Dynamic Time Warping(DTW)[1] that allows stretching in time for trajectories by replicating some points in order to get the same length. To address the noise issue, Longest Common Subsequence(LCSS)[13] adopts a robust measurement. Edit Distance on Real Sequence(ERP)[3] aims to provide a less coarse description than LCSS for distance between trajectories. Recently, Frentzos et. al propose a new similarity function based on the area between two trajectories [6]. Due to inherent different problem nature, these techniques cannot be applied to solve our problem.

Trajectory Retrieving by Locations. Here, the query is a set of locations, which aims at retrieving trajectories that are close enough to the query locations. Chen et. al study a k-BCT query, which finds k trajectories that best connect all the query locations geographically [4]. In [12], Tang et. al adopt a different distance measure and provide a robust approach to processing k-NNT query in the trajectory database. However, both of them only focus on the geographical distance between trajectory points and query locations while ignoring the road condition and traveling time information. Therefore, their techniques cannot be used to answer our k-TLT query.

6 Conclusion and Future Work

In this paper, we studied a new trajectory retrieving query, k-TLT, which aims at retrieving top-k trajectories by locations and ranked by traveling time. Different from previous works which only focus on the geographically connectivity of query location, we also considered the road condition and the traveling time information. We proposed a two-step approach to efficiently answering *k-TLT* query: candidate generation, and refine-verification. In candidate generation, we took the advantage of an augmented R-tree to retrieve trajectory points in non-decreasing order with respect their reach time to a query location, and the advantage of bitset to efficiently generate a candidate set C, which contains the results of *k-TLT* query. In refine-verification, we utilized the lower bound reach time of trajectories in C to efficiently verify the top-k trajectories. Finally, we validated the efficiency of our proposed algorithm by conducting experiments on a real trajectory dataset.

As future works, we will try to adopt more appropriate speed information in the practical context. Instead of using average speed in a certain trajectory point, we will derive the speed according to the traffic or weather conditions of the region where the certain point is located. Note the speed derived here is not only based on the single trajectories, but a result obtained from the trajectories in a corresponding small region. With these more reasonable speed information, more acceptable trajectories towards *k-TLT* query can be retrieved.

Acknowledgement. The work is supported by NSFC61232006, NSFC61021004, ARC DP120104168, ARC DP140103578 and DE120102144.

References

1. Berndt, D.J., Clifford, J.: Using dynamic time warping to find patterns in time series. In: KDD Workshop, vol. 10, pp. 359–370 (1994)
2. Chen, L., Ng, R.: On the marriage of lp-norms and edit distance. In: VLDB, pp. 792–803. VLDB Endowment (2004)
3. Chen, L., Özsu, M.T., Oria, V.: Robust and fast similarity search for moving object trajectories. In: SIGMOD, pp. 491–502. ACM (2005)
4. Chen, Z., Shen, H.T., Zhou, X., Zheng, Y., Xie, X.: Searching trajectories by locations: An efficiency study. In: SIGMOD, pp. 255–266. ACM (2010)
5. Frentzos, E., Gratsias, K., Pelekis, N., Theodoridis, Y.: Nearest neighbor search on moving object trajectories. In: Medeiros, C.B., Egenhofer, M.J., Bertino, E. (eds.) SSTD 2005. LNCS, vol. 3633, pp. 328–345. Springer, Heidelberg (2005)
6. Frentzos, E., Gratsias, K., Theodoridis, Y.: Index-based most similar trajectory search. In: ICDE, pp. 816–825. IEEE (2007)
7. Guttman, A.: R-trees: A dynamic index structure for spatial searching, vol. 14. ACM (1984)
8. Hjaltason, G.R., Samet, H.: Distance browsing in spatial databases. ACM Transactions on Database Systems (TODS) 24(2), 265–318 (1999)
9. Roussopoulos, N., Kelley, S., Vincent, F.: Nearest neighbor queries. ACM Sigmod Record 24(2), 71–79 (1995)
10. Sherkat, R., Rafiei, D.: On efficiently searching trajectories and archival data for historical similarities. Proceedings of the VLDB Endowment 1(1), 896–908 (2008)
11. Song, Z., Roussopoulos, N.: K-nearest neighbor search for moving query point. In: Jensen, C.S., Schneider, M., Seeger, B., Tsotras, V.J. (eds.) SSTD 2001. LNCS, vol. 2121, pp. 79–96. Springer, Heidelberg (2001)

12. Tang, L.-A., Zheng, Y., Xie, X., Yuan, J., Yu, X., Han, J.: Retrieving k-nearest neighboring trajectories by a set of point locations. In: Pfoser, D., Tao, Y., Mouratidis, K., Nascimento, M.A., Mokbel, M., Shekhar, S., Huang, Y. (eds.) SSTD 2011. LNCS, vol. 6849, pp. 223–241. Springer, Heidelberg (2011)
13. Vlachos, M., Kollios, G., Gunopulos, D.: Discovering similar multidimensional trajectories. In: ICDE, pp. 673–684. IEEE (2002)
14. Yuan, J., Zheng, Y., Xie, X., Sun, G.: Driving with knowledge from the physical world. In: SIGKDD, pp. 316–324. ACM (2011)
15. Yuan, J., Zheng, Y., Zhang, C., Xie, W., Xie, X., Sun, G., Huang, Y.: T-drive: Driving directions based on taxi trajectories. In: SIGSPATIAL, pp. 99–108. ACM (2010)

Comprehensive Analytics of Large Data Query Processing on Relational Database with SSDs

Keisuke Suzuki[1], Yuto Hayamizu[1], Daisaku Yokoyama[1],
Miyuki Nakano[2], and Masaru Kitsuregawa[1,3]

[1] University of Tokyo
{keisuke,haya,yokoyama,kitsure}@tkl.iis.u-tokyo.ac.jp
[2] Shibaura Institute of technology
miyuki@sic.shibaura-it.ac.jp
[3] National Institute of Informatics

Abstract. Solid-state drives (SSDs) are widely used in large data processing applications due to their higher random access throughput than HDDs and capability of parallel I/O processing. The I/O bottlenecks that HDDs on database systems face can be resolved by using SSDs because of these advantages. However, access latency on cache hierarchy may become a new bottleneck in SSD-based databases. In this study, we quantitatively analyzed the behavior of SSD-based databases by taking hashjoin operation. We found that cache misses in SSD-based databases can be decreased by reducing the hashtable size to fit into the cache. This is because the I/O cost is not increased by the high throughput of the SSDs, even though the hashjoin partition files are fragmented. We also observed that cache misses are not increased by taking a multi-hashjoin query. This is because the total size of multiple hashtables can fit into the cache size in SSD-based databases, which is in contrast to HDD-based databases, where hashtables require almost all of the available memory. Overall, our analytics clarify that the performance of multiple queries in SSD-based databases can be improved by considering data access locality of the hashjoin operation and determining the appropriate hashtable size to fit into the cache.

Keywords: RDBMS, SSD, Hashjoin, OLAP.

1 Introduction

Flash solid-state disks (SSDs) are likely to improve the I/O bottleneck of data intensive applications due to their lower latency and higher throughput than conventional hard disk drives (HDDs). They are widely used in heavy I/O workload environments as their capacity is constantly growing and the price is dropping.

SSDs offer the same block interface as HDDs, so it is easy to integrate SSDs into a storage system that enables users to access both kinds of devices transparently. We expect that the throughput of SSDs is enough to resolve the I/O bottleneck of HDDs and even maintain their bandwidth. However, another performance bottleneck may occur by fully exploiting their I/O performance in

H. Wang and M.A. Sharaf (Eds.): ADC 2014, LNCS 8506, pp. 135–146, 2014.

SSD–integrated systems. Thus, in this paper, we investigate the performance of an SSD–integrated system and show that it is insufficient to simply treat SSDs as a faster disk-they are also a key element offering a new paradigm for data intensive application performance models.

The I/O costs of conventional HDD-based database systems are often larger by an order of magnitude than memory access costs and CPU calculation costs. Therefore, I/O bandwidth limits the total performance of queries. In contrast, SSDs fill the gap between I/O costs and memory access and calculation costs, especially in the case of random I/Os. If the utilization of SSDs results in removing the I/O bottleneck, memory access and calculation costs may become a new bottleneck. That means we have to consider better utilization of computing resources such as cache and memory. We comprehensively analyzed the performance of SSD-based databases by taking a hashjoin operation that is often used in large data query processing. We then obtained the following information.

- The overall performance of a hashjoin is seriously affected by cache miss penalties. These misses can be reduced by setting a small hashtable size to fit the hashtables into the cache, but this cannot be done on HDD-based databases because using a small memory space causes fragmentation of hashtable partitions and decreases the I/O throughput. In contrast, SSDs improve the I/O throughput even though there are many fragmented hashtable partitions since they have no mechanical seek time and achieve a better performance than HDDs.
- By processing a query of multiple hashjoins, such as queries of decision support systems, cache misses are likely to increase more than with a single join query since multiple hashtables may exist at the same time and share a cache. SSD-based databases can avoid such increases by reducing the individual hashtable size enough to fit some hashtables into the cache. Thus, we have to consider data access locality of hashjoins.

The primary contributions of this paper are as follows:

- We confirm that we can shrink the size of hashtable size to obtain a good total performance of query execution in SSD-based databases.
- We confirm that the potential of improving the performance of multiple query execution by setting the appropriate memory size.

The remainder of this paper is organized as follows. Related work is presented in Section 2. Section 3 explains the behavior and expected processing cost of the hashjoin operation that is used in our analysis. Section 4 shows the basic access performance of HDDs and SSDs. In Section 5, we discuss our experimental analysis of the utilization of SSDs on large data query processing. We conclude with some final insights in Section 6.

2   Related Work

Recently, there has been much research in the area of SSD-integrated database systems. These can be roughly divided into three categories of SSD usage: buffer pool extension, indexing, and HDD-SSD mixed hybrid storage management.

Concerning buffer pool extension, Bhattacharjee et al. proposed a temperature-aware caching (TAC) schema [1,2] that monitors and obtains the statistics of the access patterns of data and then decides which data to keep in the cache on the basis of their access frequency. The FaCE system [3], proposed by Kang et al., uses the multiversion FIFO cache replacement algorithm to reduce the random write. The buffer pool extension is one of promising fields of SSD usage. However, as mentioned in [4], the buffer pool extensions are not beneficial for ad hoc large data processing queries which we focus on in this paper.

Hybrid storage management resembles the idea of caching in that it basically places frequently accessed data on SSDs and less accessed data on HDDs. Koltsidas et al. [5] detect workloads for the pages and distribute read-intensive pages on SSDs and write-intensive pages on HDDs, which overcomes the random write weakness of SSDs. The hStorage-DB [6] semantically analyzes the I/O workload of queries from execution plans. This approach enables data placing prior to query execution, and for that reason, cache filling up time and monitoring overheads are not needed. Hybrid HDD-SDD usage schemes are important for SSDs with less capacity and higher price than HDDs. Our intention is first to elaborate upon the query processing for SSD-only databases.

As for indexing, the FD-tree [7] optimizes writing performance by aggregating write requests, while the PIO B-tree [3] exploits the internal parallelism of SSDs. Indices are typically used on a scan whose data selectivity is low, while a sequential scan and hashtable are likely to be used on a high-selectivity scan. Tsirogiannis et al. [8] utilize the column-based table store to exploit the random access performance of SSDs and propose a column store database oriented hashjoin algorithm.

The studies above focus only on the I/O characteristics of SSDs. In this work, we analyze not only I/O behaviors but also the entire performance improvement of database systems and other component bottlenecks.

3   Join Operation with Hashtable

With large data processing tasks such as DSS queries, a hashtable is typically used on several database operations such as aggregation, projection, and join. We use a hashjoin operation to evaluate performance improvement by SSDs since hashjoin is one of the heaviest workload operations within databases. We clarify that the high I/O throughput of SSDs affects the entire performance of hashjoin operation from the aspect of memory access latency.

3.1   Grace Hashjoin [9] and Hybrid Hashjoin [10]

When a hashtable cannot fit into the main memory due to its size, Grace join divides the target data into partitions to fit each partition into memory and stores them on a disk.

The process of Grace hashjoin is divided in two phases: build and probe. For the sake of explanation, assume the join operation of relations R, S, and hashtables are created on S. First, in the build phase, both target relations are partitioned by the same hash function and partitions are written to disk. Next, in the probe phase, two partitions R_i and S_i ($1 \leq i \leq n$, $n = S$/memorysize), which have the same hash value, are selected to join. S_i is loaded and its hashtable created on memory and then the tuples of R_i are matched with the tuples of S_i by referring to the hashtable. This operation is repeated for each partition.

Grace hashjoin consumes memory space for only the write buffer of each partition at the build phase. Hybrid hashjoin utilizes the rest of the memory space to hold the hashtable of the first partition S_1. The memory residing partition (S_1) and the counterpart partition R_1, which have the same hash value as S_1, are processed without being stored to disk. This is how hybrid hashjoin can reduce the I/O cost of an operation with S_1 and R_1.

3.2   Processing Cost of Hashjoin

An HDD-based DBMS often uses Grace hashjoin or hybrid hashjoin on large data query processing because a vast amount of I/Os seriously decreases processing throughput. The rest of this paper deals with hybrid hashjoin algorithms.

The I/O pattern of hashjoin depends on the size of working memory (working memory means available memory space for each hashjoin operation.). This is because the partition size and the number of partitions are decided to fit a respective hashtable for partitions of S into working memory. Therefore, when the working memory space is small, many small partitions are created, which results in the generation of many fragmented partition files. Many random I/Os are invoked to access these fragmented partition files. The fragmentation causes serious I/O throughput degradation on HDDs since random I/Os are 100 - 1000 times slower than sequential I/Os. For this reason, much memory space is typically assigned for hashjoin on HDDs to avoid fragmentation. Concerning SSDs, however, the random I/Os are not slower by an order of magnitude, and therefore the fragmentation has less impact on the I/O throughput. This condition enables less memory space to be used.

A hashtable generated at the probe phase is repeatedly accessed by matching a tuple of relation R. This means that data access locality is expected and has to be considered. The size of working memory is also related to the number of cache misses of the probe phase. When a hashtable for partition S_i fits into a cache, cache misses do not occur after loading partitions. The hashtable size is limited by working memory size, so a hashtable can be fit into a cache and utilization becomes high when the working memory size is smaller than the cache size. SSDs help keep the memory size small without much I/O throughput degradation. Thus, SSDs are expected not only to improve the I/O bottleneck but also to help reduce the number of cache misses on a hashjoin.

4   Basic Performance of HDDs and SSDs

The architecture of SSDs is fundamentally different from that of HDDs. Rotating disks and moving heads to address data are the bottleneck of random accesses

Table 1. Experimental platform setup

CPU	Xeon X7560 (L3 Cache: 24 MB) @ 2.27 GHz x 4
DRAM	64 GB
Storage (SSD)	ioDrive Duo x4 (8 Logical units, Software RAID0)
Storage (HDD)	SEAGATE ST3146807FC x12 (Software RAID0)
Kernel	linux-2.6.32-220
File system	ext4

on HDDs, while SSDs are pure electronic devices so they have no seek time. Another important characteristic is that current SSDs are composed of multiple flash chips, which means they are able to process some I/Os simultaneously. We took some I/O micro-benchmark programs and measured the basic I/O behaviors of SSDs and HDDs in actual use. We then analyzed the basic performances and clarified the differences between SSDs and HDDs.

4.1   Experimental Setup

Table 1 shows the platform setup of our experiment. The interface of the SSDs is PCI Express and that of the HDDs is Fibre Channel. I/O scheduler is set to noop for both storages. The SSDs are tied up by software RAID0 with chunk size = 64 kB and use ext4 file system. The HDDs are set up in the same way.

4.2   Throughput of Sequential and Random Access

To confirm that the random access of the SSDs was superior to the HDDs, we ran micro-benchmark programs of sequential and random read I/Os.

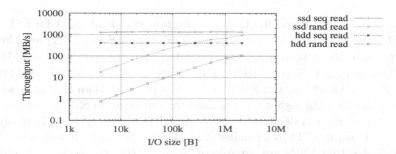

Fig. 1. I/O throughput for sequential and random accesses on SSD and HDD

Figure 1 shows the I/O throughput of SSD and HDD by varying the size of individual I/Os. The throughput of the sequential read (seq read) of SSD is 3.1 times faster than that of HDD and the throughput of the random read (rand read) is 8.6 - 23.7 times faster. The difference of I/O throughput on random read becomes larger at smaller I/O size settings. The throughput of sequential read does not depend on I/O sizes because of the read ahead function, while in contrast, the throughput of random read is proportional to I/O sizes.

4.3　Throughput of Mixture Workload

The I/O workload of actual applications consists of both read and write functions. The internal parallelism and high random access throughput of SSDs take advantage of such workloads. To demonstrate this, we used a benchmark program of a mixture of I/Os. The workload consisted of read 75% and write 25%. The ratio of read/write was similar to our experimental hashjoin operation mentioned in Section 5. The benchmark executes operations as follows. (1) Open two files: one for read operations and the other for write operations. (2) Issue I/O operations in a specified I/O size. First three read I/Os are issued, then one write I/O is issued, and then the process is repeated. The read operations sequentially scan a file and write operations add data to the other empty file.

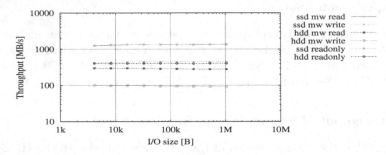

Fig. 2. I/O throughput for mixture workload of read and write accesses on SSD and HDD

Figure 2 shows the throughputs of the mixture workload (mw) for SSD and HDD. The results of readonly workloads (mentioned in Section 4.2) are plotted for comparison. Read/write throughputs of the mixture workload on SSD are both 4.2 times higher than on HDD. The difference of the read throughput of the mixture workload between SSDs and HDDs is larger than that of readonly workload. The read throughput of the mixture workload on SSD is the same as that of the readonly workload. In contrast, on HDD, the read throughput on the mixture workload is 0.74 times smaller than that of the readonly workload. This result indicates that SSDs are capable of parallel I/O processing and suitable for mixture I/O workloads.

5　Experimental Analysis of Hashjoin Operation

We performed experiments with hashjoin queries and analyzed the I/O throughput improvement and access costs on the cache hierarchy of large data query processing on an SSD-based database.

5.1   Database Setup and Workload

We used PostgreSQL [11] for RDBMS and set shared buffer size to 8 GB. We created the same databases on SSD and HDD by using data of TPC-H [12] benchmark at a scale factor of 100. Hashjoin processing performance depends on the number of partition files and hashtable size. The more the number of partition files is increased, the smaller the hashtable size becomes. In the case of HDD-based databases, it is preferable to decrease the number of partition files. This is because the fragmentation of files causes serious I/O performance degradation. There is a trade-off related to working memory size, so we handle it as a parameter to control the workload of hashjoin and observe the processing performance for each value. Work_mem is a PostgreSQL parameter that describes the in-memory buffer size per database operation, that is, the working memory size in a hashjoin. Since PostgreSQL uses hybrid hashjoin, when work_mem is larger than the entire hashtable size, no partitions are written to disk. We experimented on two queries in SSD and HDD environments: (1) a single join query, with the join part and lineitem tables on partkey, and (2) a realistic workload query, with TPC-H query 8, which contains the join of 8 tables.

We measured the query execution time by changing the work_mem size between 64 kB and 2 GB. To observe the breakdown of CPU utilization, mpstat(1) is used, and L3 cache references and cache misses are measured by a Linux profiler perf[13].

5.2   Single Join Query

To demonstrate that SSDs improve the throughput of query executions, we experimented with join operation on part and lineitem tables. Each tuple of a lineitem table is joined with one tuple of a part table which has the same partkey in this query. The table sizes of the part and the lineitem are 20 GB and 86 GB, respectively. The hashtable is created on the part table, and its total size is about 800 MB.

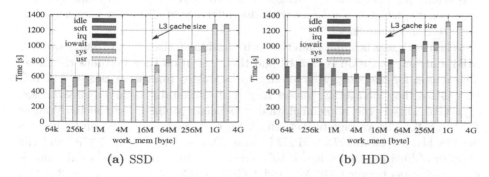

(a) SSD   (b) HDD

Fig. 3. Single join query execution time

Figure 3 shows the hashjoin execution time and its breakdown (usr, system, iowait, irq, soft irq, and idle) for the respective work_mem values in the SSD or HDD environment. In the figure, usr indicates CPU operational cost and the total of sys and iowait indicates I/O operational cost.

When work_mem is smaller than L3 cache size (64 kB - 16 MB), SSD and HDD show different trends. The smaller work_mem is, the more I/O cost is stacked up on HDD, because the I/O throughput is saturated owing to the fragmentation. In contrast, the SSD results show that I/O costs are lower than HDDs and approximately not changed in every point, which indicates remaining I/O bandwidth.

When work_mem is larger than an L3 cache size (larger than 32 MB), the CPU cost is growing in both environments and the I/O cost is no longer a bottleneck. This is due to the increased number of cache misses because the hashtable size is too big for the L3 cache size.

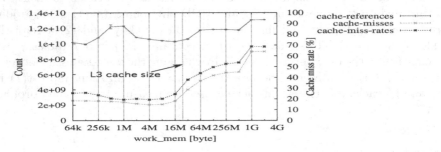

Fig. 4. Number of cache references/misses and cache miss rates for each work_mem sizes on SSD

Figure 4 shows the number of L3 cache references, misses, and miss rates for each work_mem on the SSD measurement. For example, execution with work_mem = 1 GB has 7×10^9 larger cache misses than 4 MB, and the DRAM access latency is about 100 nanoseconds in our experimental environment. Consequently, execution with work_mem = 1 GB gets a $7 \times 10^9 \times 100(ns) = 700(s)$ larger cache miss penalty, which is consistent with the difference of CPU cost between 4 MB and 1 GB in Figure 3a.

When work_mem is larger than the entire hashtable size (larger than about 800 MB), only one partition is created and then execution time becomes the same. The reason for the steeply increasing cache misses from work_mem = 512 MB to 1 GB is that the average bucket length of a hashtable is larger on work_mem = 1 GB. This is the implementation dependent problem for the hashjoin of PostgreSQL. The average bucket length for each work_mem is 2.2 on 64 kB, 3.8 on 128 kB, 5.7 on $256kB - 512MB$, and 10.5 on larger than 1 GB, and the number of lineitem tuples is 6×10^8. Then, the difference of the total number of bucket scans between 512 MB and 1 GB is $(10.5 - 5.7) \times 6 \times 10^8 \approx 3 \times 10^9$, which fits the difference of the number of cache misses in Figure 4.

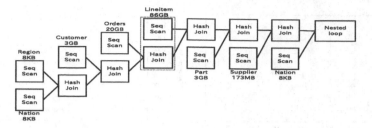

Fig. 5. The query execution plan of TPC-H query 8

5.3 TPC-H Query

We measured the TPC-H query to demonstrate that the bottleneck of query processing changes the same way as a single join in actual DSS queries. We used query 8, which contains the join of 8 tables. Some calculation parts of the query are removed, since we are interested in only the I/O performance behavior of the query. The execution plan is as shown in Figure 5. For each hashjoin operation, a hashtable is created on the bottom side node in Figure 5. At the point where I/O is the heaviest (enclosed by a red circle in Figure 5), the total hashtable size is about 400 MB.

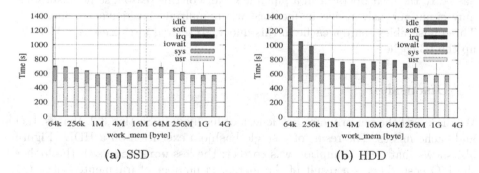

Fig. 6. TPC-H query 8 execution time

Figure 6 shows the results of query measurement on SSD and HDD. The difference of I/O cost between SSD and HDD is more conspicuous than single join query.

Figure 7 shows the I/O throughput timeline during the execution of query 8 on SSD and HDD, which are observed under work_mem = 128 kB. In the phase of the lineitem table scan (about 90 - 550 seconds in Figure 7a, 100 - 700 seconds in Figure 7b), the read I/O throughput is sometimes decreased by the write I/O, which writes hashtable partitions to storage during HDD execution. This is not observed during SSD execution, since SSDs can process multiple I/Os in parallel, as mentioned in Section 4.3. In the phase of hashjoin probing after the lineitem

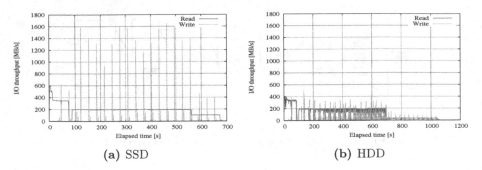

Fig. 7. The timeline of I/O throughput during query 8 execution (work_mem = 128 kB)

scan (about 550 - 680 seconds in Figure 7a, 700 - 1050 seconds in Figure 7b), the processing time on SSD is about 2.7 times faster than HDD. Since partition files are fragmented when work_mem is small, the I/O throughput of HDD is low, which inhibits the processing performance.

The I/O cost starts to increase from work_mem = 8 MB in Figure 6, even though it is smaller than the L3 cache size. In the build phase of hybrid hashjoin, in order to process the tuple matching of the memory residing hashtable without temporarily writing them to storage, the join result for those tuples is directly passed to the next operator in a pipeline style. For this reason, some hashtables may simultaneously reside in memory when multiple hashjoins are included. Three hashtables share a cache in this query, so setting work_mem as 8 MB fills up the L3 cache.

6   Discussion

We discussed the performance bottleneck of hashjoin in Section 3.2, that is, I/O and cache misses. The result of a single hashjoin execution on a HDD (Figure 3b) shows that our assumption was correct: the less work_mem used, the higher the I/O cost. This is a result of the increased number of fragmented files. On the other hand, when work_mem is larger than the L3 cache size, the CPU cost becomes large because of the increasing number of cache misses. However, the result of multiple hashjoins on an HDD (Figure 6b) suggests that the number of cache misses has no relation to the work_mem setting at a multiple hashjoin query. The CPU cost does not grow as in a single join query even if work_mem is much larger than the L3 cache size. For this reason, query execution time becomes shorter when we set work_mem to 512 MB or larger. This is a peculiar case for the high data distribution locality of lineitem and orders tables, as follows. The predicate of the third join operation, which is indicated by a red circle in Figure 5, is orders.orderkey = lineitem.orderkey. There are four tuples on average that have the same orderkey on the lineitem table. Tuples are stored in ascendant order of orderkey in the lineitem table, so several tuples with the same orderkey are concentrated in the table, that is, the data access locality occurs on a hashtable. Therefore, the same hash bucket is likely to be accessed successively at the probe phase, and consequently the number of cache misses becomes small. On the other hand, tuples with the same partkey are not concentrated, so the number of cache misses becomes large in a single query. If data locality is low in TPC-H query 8, the CPU cost increases when work_mem is larger than 8 MB, as in the results of a single query execution, and then near work_mem = 4 MB points would be optimal. In such a case, to avoid an increase of cache misses, it is inevitable to get some overheads of I/O fragmentation on the HDD-based database. These I/O cost overheads are decreased on the SSD-based database. The measurements on SSD in Figure 3a indicate that the query execution time is not affected by I/O fragmentation and rather is likely to be affected by the cache miss penalties.

Considering these results, when SSDs are used, it is better to keep the working memory size small enough to fit the hashtable into the cache. However, in the current HDD-based hashjoin implementation, very large memory is required to decrease the number of fragmented files. The working memory size can be reduced as long as the fragmentation of partitions does not cause I/O bottleneck in SSD-based databases. In our experiments, there was no I/O bottleneck even if work_mem was 64 kB.

As a result of using less memory space for a hashjoin operation, the portion of cache and memory space remains free. This remaining cache will help improve the performance of complex queries and parallel multiple queries. A complex query such as multiple hashjoin operations are executed in a pipeline manner. For example, TPC-H query 8 deploys three hashtables at the same time. If all hashtables can reside together in a cache, the number of cache misses becomes small. Another case of utilization is the parallel execution of multiple queries. The cache and other computing resources may not be fully consumed by sequential query processing. (Here, by other computing resources we mean CPU cores (most current processors have several cores internally) and I/O bandwidth (I/O bandwidth of SSDs becomes wider by parallel I/O processing such as mixture I/O workload for its internal parallelism)). Parallel query execution enables us to utilize remaining resources and improve the entire query processing performance.

7   Conclusion

In this paper, we experimentally analyzed the performance improvement and newly observed bottlenecks of large data query processing in SSD-based databases. Our experiments on hashjoin queries showed that cache miss penalties seriously affected the query processing performance. We found that it is preferable to set a small hashtable size to fit into the cache on SSD-based databases, as this reduces the number of cache misses at the probe phase. Hashtable size should be relatively large on HDD-based databases because I/O cost becomes large in a small hashtable size on HDDs. This is due to the poor I/O throughput of HDDs under fragmentation caused by generating many hashtable partition files when the hashtable is small. In contrast, the I/O cost of SSDs is not increased by the fragmentation. Thus, considering data access locality of hashjoin is more important at the query execution in SSD-based databases. Experiments on a modern SSD-based system showed that hashtable size can be reduced to 64 kB without any increase to I/O cost by the fragmentation. As a result of reducing hashtable size, the portion of cache and memory space that are not used by hashtable remains free. Those remaining resources can be utilized to improve the performance of a multiple hashjoin query such as TPC-H query 8. Another promising way to utilize the remaining resources is parallel execution of multiple queries. Exploring data access locality of multiple queries will be the focus of our future work.

Acknowledgment. This work is partially supported by JSPS KAKENHI Grant Number 24300034 and 26280130.

References

1. Bhattacharjee, B., Ross, K.A., Lang, C., Mihaila, G.A., Banikazemi, M.: Enhancing recovery using an SSD buffer pool extension. In: DaMoN 2011, pp. 10–16. ACM (2011)
2. Canim, M., Mihaila, G.A., Bhattacharjee, B., Ross, K.A., Lang, C.A.: SSD buffer-pool extensions for database systems. Proc. VLDB Endow. 1435–1446 (2010)
3. Kang, W.H., Lee, S.W., Moon, B.: Flash-based extended cache for higher through-put and faster recovery. Proc. VLDB Endow. 5(11), 1615–1626 (2012)
4. Do, J., Zhang, D., Patel, J.M., De Witt, D.J., Naughton, J.F., Halverson, A.: Turbocharging DBMS buffer pool using SSDs. In: SIGMOD 2011, pp. 1113–1124. ACM (2011)
5. Koltsidas, I., Viglas, S.D.: Flashing up the storage layer. Proc. VLDB Endow. 1(1), 514–525 (2008)
6. Luo, T., Lee, R., Mesnier, M., Chen, F., Zhang, X.: hStorage-DB: Heterogeneity-aware data management to exploit the full capability of hybrid storage systems. Proc. VLDB Endow. 5(10), 1076–1087 (2012)
7. Li, Y., He, B., Yang, R.J., Luo, Q., Yi, K.: Tree indexing on solid state drives. Proc. VLDB Endow. 3(1-2), 1195–1206 (2010)
8. Tsirogiannis, D., Harizopoulos, S., Shah, M.A., Wiener, J.L., Graefe, G.: Query Processing Techniques for Solid State Drives. In: SIGMOD 2009, pp. 59–72. ACM (2009)
9. Kitsuregawa, M., Tanaka, H., Moto-Oka, T.: Relational Algebra Machine GRACE. In: Goto, E., Furukawa, K., Nakajima, R., Nakata, I., Yonezawa, A. (eds.) RIMS 1982. LNCS, vol. 147, pp. 191–214. Springer, Heidelberg (1983)
10. Schneider, D.A., De Witt, D.J.: A performance evaluation of four parallel join algorithms in a shared-nothing multiprocessor environment. In: SIGMOD 1989, pp. 110–121. ACM (1989)
11. PostgreSQL, http://www.postgresql.org/
12. Transaction Processing Performance Council, An ad-hoc, decision support bench-mark, http://www.tpc.org/tpch/
13. Perf, https://perf.wiki.kernel.org/

Fast Information-Theoretic Agglomerative Co-clustering

Tiantian Gao and Leman Akoglu

Stony Brook University
Department of Computer Science
{tiagao,leman}@cs.stonybrook.edu

Abstract. Jointly clustering the rows and the columns of large matrices, a.k.a. co-clustering, finds numerous applications in the real world such as collaborative filtering, market-basket and micro-array data analysis, graph clustering, etc. In this paper, we formulate an information-theoretic objective cost function to solve this problem, and develop a fast *agglomerative* algorithm to optimize this objective. Our algorithm rapidly finds highly similar clusters to be merged in an iterative fashion using Locality-Sensitive Hashing. Thanks to its bottom-up nature, it also enables the analysis of the cluster hierarchies. Finally, the number of row and column clusters are automatically determined without requiring the user to choose them. Our experiments on both real and synthetic datasets show that the proposed algorithm achieves high-quality clustering solutions and scales linearly with the input matrix size.

1   Introduction

Clustering is a widely used technique that aims to group similar objects together, with numerous applications such as data summarization, classification, and outlier detection. Typically, the input data is represented as a two-mode matrix, e.g. customer-product purchasing data, document-term occurrence data, user-webpage browsing data, etc. Traditional clustering focuses only on one-mode, that is, clustering one dimension of the data matrix based on similarities along the second dimension, e.g., document clustering based on term similarity.

Another class of methods focuses on the *two-mode* clustering problem (a.k.a. bi-, co-, or block clustering), which aims at *simultaneously* clustering both dimensions of the data matrix, e.g. document clusters based on term similarity together with term clusters based on document appearance similarity. An illustration of co-clustering is given in Fig. 1. Co-clustering has many applications such as micro-array data analysis, market-basket analysis, (bi-partite) graph clustering, to name but a few. The main advantage of co-clustering is that the joint clustering of the rows and columns fully and succinctly summarizes the underlying structure of relations in the data for both types of objects.

In this paper, we propose a fast agglomerative hierarchical co-clustering technique, that scales linearly with the input matrix size. Our motivation is that agglomerative clustering techniques are known to alleviate the resolution-limit problem in clustering [10], being able to find smaller size clusters effectively. Our proposed algorithm, called CoClusLSH, rapidly finds the most similar objects

H. Wang and M.A. Sharaf (Eds.): ADC 2014, LNCS 8506, pp. 147–159, 2014.

to merge, using ideas from Locality-Sensitive Hashing, and iteratively builds the row and column cluster hierarchies. The two hierarchies are built in an alternating fashion, such that the clustering of both object types is intertwined.

In clustering, one of the main challenges is to determine the "correct" or a "good" number of clusters. In hierarchical clustering, this challenge translates to picking a level of the hierarchy to "cut", the subtrees of which determine the final clustering. It is often a hard task for the user to specify the number of output clusters, especially for large datasets. We circumvent this challenge by formulating an information-theoretic co-clustering objective cost function, based on the number of bits needed to encode the input matrix. Our goal then is to find a clustering that achieves as low of a cost as possible. We update this cost while growing up our cluster hierarchies and merge wo clusters only when it yields a lower cost. This principled way of building the clustering is exactly what guides us in "when to stop"—stop growing the hierarchies when no further merges can reduce the objective cost. As such, the number of sub-hierarchies at algorithm termination automatically gives us the number of row and column clusters.

The main merits of our method over the (cited) previous proposals are that (*i*) it automatically finds a good number of clusters [9,19], (*ii*) achieves linear scalability [12], and (*iii*) provides the cluster hierarchies [7] (see §4 for details). None of the previous approaches exhibits all three properties at the same time. We summarize our main contributions as follows:

- We propose a new technique for agglomerative co-clustering, and formulate an information-theoretic objective that enables us to determine the number of row/column clusters automatically in a principled data-driven way,
- We develop a fast algorithm called CoClusLSH that rapidly finds similar clusters to merge in order to grow the row/column hierarchies,
- We show that CoClusLSH scales linearly with the input matrix size,
- Experiments on synthetic and real datasets with ground truth cluster labels demonstrate the effectiveness and efficiency of our method.

2   Proposed Method

2.1   Problem Definition

We consider the problem of *co-clustering*, i.e. joint clustering the rows and columns of a large binary matrix (such a matrix can be thought of as a bipartite graph). In particular, given a bipartite graph with n type-1 nodes, m type-2 nodes, and their binary connectivity information, our goal is to cluster the type-1 nodes into k, and the type-2 nodes into l disjoint clusters such that the nodes in

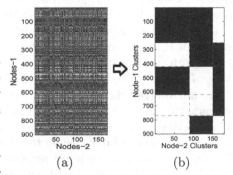

(a)   (b)

Fig. 1. (a) Example graph with n=900, and m=180, where (b) CoClusLSH finds k=5 type-1 and l=3 type-2 clusters.

the same cluster have "similar connectivity". Intuitively, a set of nodes have similar connectivity if their neighbors "highly" overlap (for e.g., see Fig. 1).

Given the above problem description, two main questions arise: (P1) how to choose the number of node clusters k and l?, and (P2) how to assign the nodes to their "proper" clusters? (P2) aims at summarizing the adjacency matrix $\mathbf{A}$ of the graph with homogeneous, rectangular regions of high and low densities, while (P1) deals with choosing the right number of clusters and hence the number of these rectangular regions. Roughly speaking, having more clusters allows us to obtain more homogeneous regions. At the very extreme we can have $n \times m$ "regions" each with perfect 0 or 1 density which, however, does not provide any summary. As such, a co-clustering algorithm should achieve a good trade-off between homogeneity and the number of regions. Intuitively, this trade-off calls for model selection, which brings us to the next section.

2.2   Problem Formulation

In order to achieve a proper balance between the homogeneity and the number of the rectangular regions, we use a similar objective function to [7] founded on the Minimum Description Length (MDL) principle [25]. MDL provides a model selection criterion based on lossless compression principles, where the objective is to compress/transmit/store the adjacency matrix $\mathbf{A}$ using as few bits as possible. The compression cost consists of two main parts: the number of bits required to encode (1) the clustering "summary" (model description cost), and (2) each rectangular region (data description cost) given the model.

Next we describe each part in detail in the context of our objective function after providing the notation.

Notation. Let k and l respectively denote the number of disjoint row- and column-clusters, and $R : \{1, 2, \ldots, n\} \rightarrow \{1, 2, \ldots, k\}$ and $C : \{1, 2, \ldots, m\} \rightarrow \{1, 2, \ldots, l\}$ denote the assignments of rows to row-clusters and columns to column-clusters. We refer to (R, C) as a *mapping*. To better describe a mapping, let us rearrange the rows and columns of the adjacency matrix $\mathbf{A}$ such that all rows corresponding to row-cluster-1 are listed first, followed by rows in row-cluster-2, and so on. We also rearrange the columns in a similar fashion using column-cluster assignments. One can imagine that such a rearrangement sub-divides $\mathbf{A}$ into $k \times l$ two-dimensional, rectangular blocks (as in Fig. 1 (b)), which we will refer to as B_{ij}, $i = 1, \ldots, k$ and $j = 1, \ldots, l$. Finally, let (r_i, c_j) denote the dimensions of B_{ij}, where r_i denotes the size of row cluster i, and c_j denotes the size of column cluster j.

Objective Function. Our objective function consists of a two-part (lossless) compression cost of the adjacency matrix $\mathbf{A}$. This cost can be thought of as the total number of bits required to encode $\mathbf{A}$. The first part is the model description cost that consists of describing the mapping (R, C). The second part is the data description cost that consists of encoding the sub-matrices (i.e., the B_{ij} "blocks"), given the mapping. Intuitively, a good choice of (R, C) would compress $\mathbf{A}$ well, and yield a low total description cost. In particular:

The *Model Description Cost* consists of encoding the number of row and column clusters and the corresponding mapping.

- The matrix dimensions of $\mathbf{A}$ require $\log^* n + \log^* m$ bits, where $\log^*$ denotes the universal code length for integers.[1] This term is independent of any particular mapping.
- The number of row and column clusters (k, l) require $\log^* k + \log^* l$ bits.
- The row and column cluster assignments with arithmetic coding require $nH(P) + mH(Q)$ bits, where H denotes the Shannon entropy function, P is a multinomial random variable with the probability $p_i = \frac{r_i}{n}$ and r_i is the size of the i-th row cluster, $1 \leq i \leq k$. Similarly, Q is another multinomial random variable with the probability $q_j = \frac{c_j}{m}$ and c_j is the size of the j-th column cluster, $1 \leq j \leq l$.

The *Data Description Cost* consists of encoding the actual blocks.

- For each block B_{ij}, $i = 1, \ldots, k$, $j = 1, \ldots, l$, encoding $n_1(B_{ij})$, i.e. the number of 1s it contains, takes $\log_2(r_i c_j + 1)$ bits.
- To encode the actual blocks B_{ij}, we first calculate their density $P_{ij}(1) = n_1(B_{ij})/n(B_{ij})$, where $n(B_{ij}) = n_1(B_{ij}) + n_0(B_{ij}) = r_i c_j$. Then, the number of bits required to encode each block can be written as:

$$E(B_{ij}) = -n_1(B_{ij})\log_2(P_{ij}(1)) - n_0(B_{ij})\log_2(P_{ij}(0)) = n(B_{ij})H(P_{ij}(1)).$$

Overall, the *Total Encoding Cost (Length $L(\mathbf{A}; R, C)$ in bits)* becomes

$$L(\mathbf{A}; R, C) = \log^* n + \log^* m + \log^* k + \log^* l +$$

$$\sum_{i=1}^{k} r_i \log_2(\frac{n}{r_i}) + \sum_{j=1}^{l} c_j \log_2(\frac{m}{c_j}) + \sum_{i=1}^{k}\sum_{j=1}^{l}\left(\log_2(r_i c_j + 1) + E(B_{ij})\right) \quad (1)$$

2.3   Proposed Algorithm CoClusLSH

Minimizing our objective function in Equ. (1) is intractable for very large graphs as the number of possible orderings of rows/columns is combinatorial. Thus, we develop an algorithm that aims at finding a fast approximate solution.

Our CoClusLSH algorithm starts by assigning each row and column in $\mathbf{A}$ to their respective clusters. In the main loop, it alternates between trying to merge candidate column and row clusters, for reduced cost. In order to rapidly find sufficiently similar candidate clusters, it employs the LSH technique [11] and generates a *signature* for each cluster which is then used to hash the clusters into multiple hash tables. Candidate clusters hashed to the same buckets are then tested for merge. The algorithm terminates when no more merges can be done for lower cost. The clusters at termination constitutes the final set of clusters, and the intermediate merge operations define the cluster hierarchies.

We provide the detailed pseudocode for CoClusLSH in Algorithm 1.

---

[1] The optimal number of bits required to encode a positive integer x whose range is unknown is $\log^* x \approx \log_2 x + \log_2 \log_2 x + \ldots$ of the positive terms [25].

Algorithm 1. CoClusLSH

Input: $n \times m$ adjacency matrix $\mathbf{A}$, LSH parameters r, b
Output: A heuristic solution towards minimizing total encoding $L(\mathbf{A}; R, C)$:
number of row and column groups (k^*, l^*), associated mapping (R^*, C^*)

1. Set $R^0 := \{1, 2, \dots, n\} \to \{1, 2, \dots, n\}$ Set $C^0 := \{1, 2, \dots, m\} \to \{1, 2, \dots, m\}$

2. Set $k^0 = n$, $l^0 = m$. Let T denote the outer iteration index. Set $T = 0$.
3. **repeat**
4. $C^{T+1}, l^{T+1} :=$ MERGE-COLCLUS$(\mathbf{A}, C^T, l^T, k^T, r, b)$
5. $R^{T+1}, k^{T+1} :=$ MERGE-ROWCLUS$(\mathbf{A}, R^T, k^T, l^T, r, b)$
6.   **if** $L(\mathbf{A}; R^{T+1}, C^{T+1}) \geq L(\mathbf{A}; R^T, C^T)$ **then**
7.   **return** $(k^*, l^*) = (k^T, l^T)$, $(R^*, C^*) = (R^T, C^T)$
8.   **else** Set $T = T + 1$ **end if**
9. **until** convergence

Procedure 1. MERGE-COLCLUS (**Procedure 2** MERGE-ROWCLUS is similar)

Input: $n \times m$ adjacency matrix $\mathbf{A}$, C^T, l^T, k^T, LSH parameters r, b
Output: C^{T+1}, l^{T+1}

1. {**Step 1. Generate signatures**} initialize signature matrix $S[i][j] \in \mathbb{R}^{rb \times l^T}$
2. **if** $T = 0$ **then** {use Jaccard similarity // *generate min-hash signatures* }
3.   **for** $i = 1$ to rb **do**
4. $\pi_i \leftarrow$ generate random permutation $(1 \dots n)$
5.   **for** $j = 1$ to l^T **do** $S[i][j] \leftarrow min_{v \in N_j} \pi_i(v)$ **end for**
6.   **end for**
7. **else** {use cosine similarity //*generate random-projection signatures*}
8.   **for** $i = 1$ to rb **do**
9. $rnd_i \leftarrow$ pick a random hyperplane $\in \mathbb{R}^{k^T \times 1}$
10.   **for** $j = 1$ to l^T **do** $S[i][j] \leftarrow sign(P_{\cdot j}(1) \cdot rnd_i)$ **end for**
11.   **end for**
12. **end if**
13. {**Step 2. Generate hash tables**}
14. **for** $h = 1$ to b **do**
15.   **for** $j = 1$ to l^T **do** $hash(S[(h-1)r + 1 : hr][j])$ **end for**
16. **end for**
17. {**Step 3. Merge clusters from hash tables** }
18. Build candidate groups: union of elements that hash to *at least one* same
    bucket in all hash tables, i.e. $c_1, c_2 \in g$ if $hash_h(c_1) = hash_h(c_2)$ for $\exists h$.
19. **for** each each candidate group g **do**
20.   **while** more merges happen **do**
21. $c_r \leftarrow$ pick a random element (col. cluster) from g
22.   **for all** clusters $c \in g$, $C^T(c) \neq C^T(c_r)$ **do**
23. $L^U \leftarrow$ update cost when $C^T(c)$ and $C^T(c_r)$ are merged by Equ. (2)
24.   **if** $L^U < 0$ **then**
25. $l^T = l^T - 1$. $C^T(c) = C^T(c_r) = min(C^T(c), C^T(c_r))$.

26.   Merge B_{ic} and B_{ic_r} $\forall i,\ 1 \leq i \leq k^T$.
27.   **end if**
28.   **end for**
29.   **end while**
30. **end for**
31. $l^{T+1} = l^T,\ C^{T+1} = C^T$

Algorithm Details. MERGE-COLCLUS, and similarly MERGE-ROWCLUS, consists of three main steps: (1) generate LSH signatures (Line 1), (2) generate hash tables (Line 13), and (3) merge clusters using hash buckets (Line 17).

In the first iteration of MERGE-COLCLUS, i.e. $T = 0$, the clusters consist of singleton nodes. As the similarity measure, we use Jaccard similarity which is high for those nodes with many exclusive common neighbors. Min-hashing is designed to capture the Jaccard similarity between binary vectors (Lines 2-6). For $T > 0$, the clusters consist of multiple nodes. As the similarity measure of two (column) clusters c_1 and c_2, we use the density similarity of their corresponding row blocks B_{ic_1} and B_{ic_2}, $\forall i$. As such, each column cluster c can be represented by a length k^T vector in which the entries denote the density $P_{ic}(1)$ of each row block i. We use their cosine similarity to compare two real-value vectors. To capture cosine similarity, we generate random-projection-based signatures (Lines 7-12). At the end of step (1), each cluster has a length-rb signature.

In step (2), we split the signature of each cluster into b length-r sub-signatures, and hash each sub-signature using standard hashing (Lines 14-16).

Step (3) involves the main merging operations. First we construct the group of candidate clusters to be merged. We put all clusters that hash to the *same* hash bucket in *at least one* hash table into the same group (Line 18). Next, we iterate over the groups to identify those clusters the merge of which will reduce the total cost (Line 19). We pick a cluster at random from a given group and test it against other clusters in the group, where we merge two clusters if the cost reduces. We continue the merges until no more merges can be done for lower cost (Lines 20-30). By focusing only on the highly similar candidate clusters within groups, MERGE-COLCLUS omits the consideration of merge between all clusters; this contributes to a reduction in the running time while enabling the merge among good candidate clusters that are highly similar.

A crucial computation in step (3) is to update the total cost when two candidate clusters are merged (Line 23). In the following we show that the update-cost L^U can be computed *locally* without requiring the re-computation of the total cost. As such, we decide to merge two clusters if their update-cost is less than 0 (Line 24), i.e. when the merge *reduces* the total objective cost.

Updating the Total Objective Cost. When two column (or row) clusters are merged, we can analyze how the encoding cost is expected to change. Without loss of generality, assume two column clusters of *sizes* c_1 and c_2 are to be merged.

Lemma 1. *If two clusters are merged, then the total cluster assignment cost, i.e.* $\sum_{j=1}^{l} c_j \log_2(\frac{m}{c_j})$, *will decrease.*

Proof. The assignment cost $c_j \log_2 \frac{m}{c_j}$ remains the same for $c_j \neq c_1, c_2$. We have

$$(c_1 + c_2) \log_2 \frac{m}{(c_1 + c_2)} = c_1 \log_2 m + c_2 \log_2 m - c_1 \log_2(c_1 + c_2) - c_2 \log_2(c_1 + c_2)$$

$$< c_1 \log_2 m + c_2 \log_2 m - c_1 \log_2 c_1 - c_2 \log_2 c_2 = c_1 \log_2 \frac{m}{c_1} + c_2 \log_2 \frac{m}{c_2}. \qquad \square$$

Lemma 2. *If two clusters are merged, then total cost* $\sum_{i=1}^{k} \sum_{j=1}^{l} \log_2(r_i c_j + 1)$ *of encoding the number of 1s for blocks will decrease.*

Proof. The $\log_2(r_i c_j + 1)$ cost remains the same for clusters $c_j \neq c_1, c_2, \forall i$.
$$\log_2(r_i(c_1 + c_2) + 1) = \log_2(r_i c_1 + r_i c_2 + 1) < \log_2(r_i^2 c_1 c_2 + r_i c_1 + r_i c_2 + 1)$$
$$= \log_2((r_i c_1 + 1)(r_i c_2 + 1)) = \log_2(r_i c_1 + 1) + \log_2(r_i c_2 + 1).$$

Lemma 3. *When two clusters merge, block encoding cost* $\sum_{i=1}^{k} \sum_{j=1}^{l} E(B_{ij})$ *will increase, i.e., if* $B_i = [B_{i1} B_{i2}]$, *then* $E(B_{i1}) + E(B_{i2}) \leq E(B_i), \forall i$.

Proof. $E(B_{ij})$ remains the same for clusters $c_j \neq c_1, c_2, \forall i$. We have $\forall i$,

$$E(B_i) = n(B_i)H\left(\frac{n_1(B_i)}{n(B_i)}\right) = n(B_i)H\left(\frac{n(B_{i1})P_{B_{i1}}(1) + n(B_{i2})P_{B_{i2}}(1)}{n(B_i)}\right)$$

$$\geq n(B_{i1})H(P_{B_{i1}}(1)) + n(B_{i2})H(P_{B_{i2}}(1)) = E(B_{i1}) + E(B_{i2})$$

where the inequality follows from the concavity of the entropy function $H(\cdot)$. (also note that $n(B_{i1}) + n(B_{i2}) = n(B_i)$). $\qquad \square$

Overall, the difference between the increase in the block encoding cost (Lemma 3) and the decrease in the cluster assignment and number of non-zeros encoding costs (Lemma 1 & 2) will determine whether two candidate clusters are merged or not (in Line 23 of Procedure 1). This difference can be computed quickly without requiring the re-computation of the total cost. Specifically, the total update-cost L^U when two (column) clusters of sizes c_1 and c_2 are merged (where there are totally m columns, and l column clusters) is equal to

$$L^U = -c_1 \log_2 \frac{m}{c_1} - c_2 \log_2 \frac{m}{c_2} + (c_1 + c_2) \log_2 \frac{m}{(c_1 + c_2)}$$

$$\sum_i -\log_2(r_i c_1 + 1) - \log_2(r_i c_2 + 1) + \log_2(r_i(c_1 + c_2) + 1)$$

$$\sum_i \left(-E(B_{i1}) - E(B_{i2}) + E(B_i)\right) - \log^\star(l) + \log^\star(l - 1) \qquad (2)$$

where the last two terms account for the difference in encoding cost of the number of column clusters, which will reduce by 1 in case of a merge.

Finally, we provide the time complexity for all steps of our method in Table 1. Details are omitted for lack of space.

Table 1. Computational complexity of CoClusLSH steps

	$T = 0$	$T \geq 1$
Step 1.	$O(n_1(\mathbf{A})rb)$	$O(klrb)$
Step 2.	$O((n + m)rb)$	$O((k + l)rb)$
Step 3.	$O((n + m)M)$	$O((k + l)M)$

3   Experiments

We next evaluate our method based on (1) clustering quality, and (2) scalability. on both real and synthetic datasets (with/without ground truth cluster labels).

3.1 Synthetic Datasets

Data Generation. To create synthetic data matrices, we use two different schemes. In the first scheme, we fix a cluster size s, and increase the number of such clusters k to obtain larger and larger graphs. In the second scheme, we fix the number of clusters, and increase the size of the clusters to obtain various size graphs. The generated matrices are diagonally strong, i.e. the density of the diagonal blocks is p, whereas the off-diagonal blocks are all-zeros. Next, we add random noise by adding ϵ fraction of non-zeros in the original matrix at random entries, to study the effect of varying noise levels on the clustering performance.

We call the first set of synthetic graphs as CAVE1 graphs, with $s = 500$, $2 \leq k \leq 11$, $p = 0.9$, and $\epsilon = \{0.1, 0.2, 0.3, 0.4\}$. This gives us 4 sets of 10 graphs of various sizes, where each set of graphs have a different level of noise.

The second set of graphs are called CAVE2 graphs, with $s = 50t, 1 \leq t \leq 10$, $k = 10$, and p and ϵ as before. This way we also obtain 40 graphs. We provide statistics of the largest CAVE1 and CAVE2 graph with 40% noise in Table 2.

Clustering Quality. We first study the effect of noise on the clustering quality. As we work with synthetic datasets, we have the ground truth for the cluster assignments and thus can compute the true/optimal encoding costs.

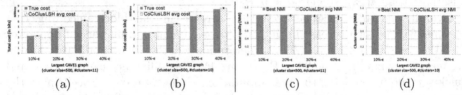

(a)   (b)   (c)   (d)

Fig. 2. (a,b) True vs. CoClusLSH cost (y-axes), (c,d) Optimal vs. CoClusLSH NMI (y-axes). (both avg. over 10 runs, bars depict σ) on the largest (a,c) CAVE1 and (b,d) CAVE2 graph with varying noise levels 10%-40% (x-axes).

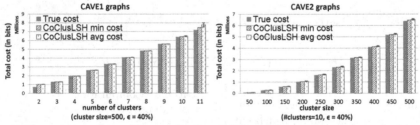

Fig. 3. True cost vs. best and avg. CoClusLSH cost (over 10 runs, bars depict σ) on all (left) CAVE1 and (right) CAVE2 graphs, when $\epsilon = 40\%$

In Figure 2 (a,b) we show the true cost in comparison to our CoClusLSH's cost for the largest CAVE1 and CAVE2 graphs, for the increasing noise levels. We observe that the gap between the optimal and CoClusLSH's cost increases with more noise, however CoClusLSH still finds good approximate solutions.

To assess the cluster assignment quality, we use the Normalized Mutual Information (NMI), a widely used measure for evaluating the clustering accuracy of a method against the ground truth clustering [18]. The ideal NMI score is 1.

Table 2. Datasets used in this work

Dataset	$\mathbf{A}$ Dim. $n \times m$	$n_1(\mathbf{A})$
US-SENATE	108 senators $\times$ 696 bills	40,609
US-HOUSE	451 rep.s $\times$ 1,646 bills	501,602
POLBLOGS	362 blogs $\times$ 5,895 words	776,870
DBLP	1,230 papers $\times$ 1,230 papers	19,267
CLASSIC	3,893 doc.s $\times$ 4,303 words	176,347
NIPS	10,617 words $\times$ 2,864 authors	160,059
YOUTUBE	77,381 users $\times$ 30,087 groups	260,240
CAVE1–40%ϵ	5,500 $\times$ 5,500	3,289,021
CAVE2–40%ϵ	5,000 $\times$ 5,000	2,974,778

Figure 2 (c,d) depicts CoCLusLSH's average NMI score for the largest CAVE1 and CAVE2 graphs, for increasing noise. As before, we observe that NMI drops slightly with more noise, while it remains > 0.9 at all noise levels.

Next, we study the effect of increasing number of clusters (as in CAVE1 graphs) and of increasing cluster sizes (as in CAVE2 graphs) on the performance. Figure 3 shows the true encoding cost against CoCLusLSH's best and average cost across all CAVE1 and CAVE2 graphs, respectively, for the most challenging case of $\epsilon = 40\%$. We observe that CoCLusLSH recovers a low cost solution in all cases.

Scalability. Next, we study the growth in the running time of CoCLusLSH with increasing graph size. In §2.3 we showed that the running time is proportional to the number of nonzeros, the number of rows and columns, and the

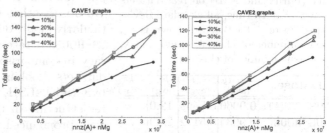

Fig. 4. Total running time (in sec) of our CoCLusLSH on (left) CAVE1 and (right) CAVE2 graphs with varying noise

maximum number of clusters that hash to the same group. In Figure 4, we show the total time w.r.t. these parameters, for all CAVE1 and CAVE2 graphs, at varying noise levels. We observe that the run time grows linearly, and that more noise demands more time for our algorithm to converge.

3.2   Real Datasets

Data description. Our real-world datasets include (Table 2): US-SENATE with senators and US-HOUSE with The House representatives voting (1 'yes', 0 'no') on congressional bills in the 111th US Congress; POLBLOGS with political blogs and the words they use; DBLP with academic papers and their commonly used terms relations; and finally CLASSIC with documents and the words they contain.

For the five datasets described above, we have the ground truth labels but only for the rows. In particular US-SENATE and US-HOUSE both consist of two

156   T. Gao and L. Akoglu

classes; (1) liberal and (2) conservative congressmen, POLBLOGS also has two classes; (1) liberal and (2) conservative blogs, DBLP contains papers from four classes of venues; (1) SIGIR (information retrieval), (2) STOC+FOCS (theory), (3) AAAI (artificial intelligence), and (4) TODS (database systems), and finally CLASSIC consists of documents from three different classes; (1) MEDLINE (medical journals), (2) CISI (information retrieval) and (3) CRANFIELD (aero-dynamics). In addition, we used two more real datasets, namely NIPS and YOUTUBE, with many rows and columns for our scalability experiments. Unfortunately, there are no class labels for the rows or columns for these datasets.

Clustering Quality. We evaluate CoClusLSH's clustering performance using two clustering quality measures, purity and NMI as before [18], using the ground truth labels of the row nodes (note that we do not have the ground truth labels for the column nodes for our datasets, thus we cannot compute the optimal encoding cost as for the synthetic datasets).

Purity measures the coherence of labels within each cluster. It, however, is expected to increase with the number of output clusters—in the extreme case where each node belongs to its own cluster, purity becomes 1. NMI can trade off the quality of the clustering against the number of clusters.

Table 3. CoClusLSH clustering quality (purity and NMI) on real datasets (avg. over 10 runs, all standard deviations were < 0.03). Also shown is k^*, avg. number of clusters found and k, the true number of clusters.

Dataset	Purity (avg)	NMI (avg)	k^* (avg)	k (true)
US-SENATE	0.9960	0.5569	12.0	2
US-HOUSE	0.9934	0.5172	28.9	2
POLBLOGS	0.5539	0.0142	18.9	2
DBLP	0.4723	0.0949	6.6	4
CLASSIC	0.3987	0.0241	5.0	3

We report both measures and the number of clusters CoClusLSH outputs on our real datasets in Table 3. We observe that CoClusLSH does particularly well on the US-SENATE and US-HOUSE datasets, while clustering accuracy in comparison is lower for the other datasets. Looking at the output clusters by CoClusLSH, we realize that the cluster structure in the congress datasets is well pronounced, while the rest of the real datasets are quite sparse with no clear cluster structure. We show an example output of CoClusLSH on the US-HOUSE and DBLP datasets in Figure 5, where CoClusLSH performs well on recovering salient cluster structure.

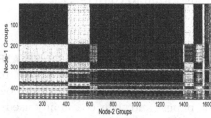 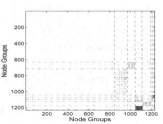

Fig. 5. Adjacency matrix of (left) US-HOUSE and (right) DBLP, with rows and columns arranged by the cluster assignments of CoClusLSH.

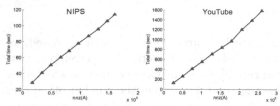

Fig. 6. Total running time of CoCLusLSH on growing (left) NIPS and (right) YOUTUBE graphs.

Scalability. Finally, we also experimentally study that the running time of CoCLusLSH with respect to the input size on real graphs. For running time measurements, we use the NIPS and YOUTUBE datasets with many rows each. To generate graphs of growing size, we increasingly sample the rows of these matrices, and report average running time over 10 runs in Figure 6. We observe that the running time grows linearly with the input graph size.

4   Related Work

Clustering algorithms in the row-mode *only* include k-means and its parameter-free variants [13,23], spectral [21], and (probabilistic) hierarchical clustering [15,20,27]. Our problem deals with *simultaneous* clustering of rows and columns, known as bi(dimensional)-, co-, or block clustering. Information-theoretic co-clustering [9] employs a lossy coding scheme to co-cluster a two-dimensional joint probability distribution, however, it requires the number of clusters as input. Conjunctive clustering [19] finds top-k largest bi-clusters in bipartite graphs, but requires two parameters for lower-bounding cluster sizes in each dimension, a diversity parameter controlling overlap, and a density parameter.

Hierarchical tiling [12] extracts nested tiles (or blocks) of various densities in the adjacency matrix. While parameter-free, it involves a quadratic preprocessing step for row/column re-ordering. Bi-clustering has also been explored in bioinformatics [1,24] often requiring the number of clusters as user input.

Cross-associations [7] automatically selects the number of clusters and scales well to large graphs. It has been used in graph partitioning [6], and extended to time-evolving [28] and attributed graphs [4]. Our approach exhibits the same merits as [7], while enabling the cluster hierarchies.

Other relevant work to ours include frequent itemset and association rules mining [3,5,14], where the support parameter is critical. In information retrieval LSI [8] uses SVD to find latent concepts in the data matrix, which requires the number of hidden concepts. In addition, subspace clustering [2,17] aims at finding all the dense clusters in all subspaces. These methods often take input parameters such as density and size thresholds to quickly scan their search space (also see [16] for a survey on subspace, correlation, and pattern-based clustering).

Finally, LSH has been used to accelerate several other problems, such as similarity search [26], outlier detection [29], and k-nearest neighbor search [22].

5   Conclusion

We propose an approach for co-clustering large binary matrices, based on an information-theoretic objective. To solve our objective, we develop a fast, bottom-up clustering algorithm that rapidly finds the most similar rows/columns to merge using locality-sensitive hashing, and determines the number of clusters automatically. We demonstrate the effectiveness and scalability of the proposed approach on real and synthetic datasets, where our algorithm recovers the cluster structure with high accuracy, and has a running time that grows linearly with the input matrix size. Our source code is freely available for academic use.[2]

Acknowledgements. This material is based upon work supported by the Army Research Office under Contract No. W911NF-14-1-0029 and Stony Brook University Office of Vice President for Research. Any findings and conclusions expressed in this material are those of the author(s) and do not necessarily reflect the views of the funding parties.

References

1. Abdullah, A., Hussain, A.: A new biclustering technique based on crossing minimization. Neurocomputing 69(16-18), 1882–1896 (2006)
2. Agrawal, R., Gehrke, J., Gunopulos, D., Raghavan, P.: Automatic subspace clustering of high dimensional data. In: SIGMOD, pp. 94–105 (1998)
3. Agrawal, R., Imieliński, T., Swami, A.: Mining association rules between sets of items in large databases. SIGMOD 22(2), 207–216 (1993)
4. Akoglu, L., Tong, H., Meeder, B., Faloutsos, C.: Pics: Parameter-free identification of cohesive subgroups in large attributed graphs. In: SDM (2012)
5. Calders, T., Goethals, B.: Mining all non-derivable frequent itemsets. In: Elomaa, T., Mannila, H., Toivonen, H. (eds.) PKDD 2002. LNCS (LNAI), vol. 2431, pp. 74–86. Springer, Heidelberg (2002)
6. Chakrabarti, D.: AutoPart: Parameter-free graph partitioning and outlier detection. In: Boulicaut, J.-F., Esposito, F., Giannotti, F., Pedreschi, D. (eds.) PKDD 2004. LNCS (LNAI), vol. 3202, pp. 112–124. Springer, Heidelberg (2004)
7. Chakrabarti, D., Papadimitriou, S., Modha, D.S., Faloutsos, C.: Fully automatic cross-associations. In: ACM SIGKDD, pp. 79–88 (2004)
8. Deerwester, S., Dumais, S., Furnas, G., Landauer, T., Harshman, R.: Indexing by latent semantic analysis. JASI 41(6), 391–407 (1990)
9. Dhillon, I., Mallela, S., Modha, D.: Information- theoretic co-clustering. In: ACM SIGKDD (2003)
10. Fortunato, S., Barthélemy, M.: PNAS, 104(1), 36 (2007)
11. Gionis, A., Indyk, P., Motwani, R.: Similarity search in high dimensions via hashing. In: VLDB, pp. 518–529 (1999)
12. Gionis, A., Mannila, H., Seppänen, J.K.: Geometric and combinatorial tiles in 0-1 data. In: Boulicaut, J.-F., Esposito, F., Giannotti, F., Pedreschi, D. (eds.) PKDD 2004. LNCS (LNAI), vol. 3202, pp. 173–184. Springer, Heidelberg (2004)

---

[2] COCLUSLSH code: http://www.cs.stonybrook.edu/~leman/pubs.html#code

13. Hamerly, G., Elkan, C.: Learning the k in k-means. In: NIPS (2003)
14. Han, J., Wang, J., Lu, Y., Tzvetkov, P.: Mining top-k frequent closed patterns without minimum support. In: ICDM, pp. 211–218 (2002)
15. Karypis, G., Han, E.-H., Kumar, V.: Chameleon: Hierarchical clustering using dynamic modeling. IEEE Computer 32(8) (1999)
16. Kriegel, H.-P., Kröger, P., Zimek, A.: Clustering high-dimensional data: A survey. TKDD 3(1), 1:1–1:58 (2009)
17. Kröger, P., Kriegel, H.-P., Kailing, K.: Density-connected subspace clustering for high-dimensional data. In: SDM (2004)
18. Manning, C.D., Raghavan, P., Schtze, H.: Introduction to Information Retrieval. Cambridge University Press (2008)
19. Mishra, N., Ron, D., Swaminathan, R.: On finding large conjunctive clusters. In: Schölkopf, B., Warmuth, M.K. (eds.) COLT/Kernel 2003. LNCS (LNAI), vol. 2777, pp. 448–462. Springer, Heidelberg (2003)
20. Newman, M.E.J.: Fast algorithm for detecting community structure in networks. Physical Review E 69 (2004)
21. Ng, A.Y., Jordan, M.I., Weiss, Y.: On spectral clustering: Analysis and an algorithm. In: NIPS (2001)
22. Pan, J., Manocha, D.: Bi-level locality sensitive hashing for k-nearest neighbor computation. In: ICDE, pp. 378–389 (2012)
23. Pelleg, D., Moore, A.: X-means: Extending K-means with efficient estimation of the number of clusters. In: ICML (2000)
24. Reiss, D.J., Baliga, N.S., Bonneau, R.: Integrated biclustering of heterogeneous genome-wide datasets. BMC Bioinformatics 7, 280 (2006)
25. Rissanen, J.: A universal prior for integers and estimation by minimum description length. The Annals of Statistics 11(2), 416–431 (1983)
26. Satuluri, V., Parthasarathy, S.: Bayesian locality sensitive hashing for fast similarity search. PVLDB 5(5), 430–441 (2012)
27. Slonim, N., Tishby, N.: Agglomerative information bottleneck. In: NIPS (1999)
28. Sun, J., Faloutsos, C., Papadimitriou, S., Yu, P.S.: Graphscope: parameter-free mining of large time-evolving graphs. In: ACM SIGKDD, pp. 687–696 (2007)
29. Wang, Y., Parthasarathy, S., Tatikonda, S.: Locality sensitive outlier detection: A ranking driven approach. In: ICDE, pp. 410–421 (2011)

Semi-supervised Learning for Cyberbullying Detection in Social Networks

Vinita Nahar[1,*], Sanad Al-Maskari[1], Xue Li[1], and Chaoyi Pang[2]

[1] School of Information Technology and Electrical Engineering,
The University of Queensland, Australia
{v.nahar,s.almaskari}@uq.edu.au, xueli@itee.uq.edu.au,
[2] The Australian E-Health Research Center, CSIRO, Australia
Chaoyi.Pang@csiro.au

Abstract. Current approaches on cyberbullying detection are mostly static: they are unable to handle noisy, imbalanced or streaming data efficiently. Existing studies on cyberbullying detection are mainly supervised learning approaches, assuming data is sufficiently pre-labelled. However this is impractical in the real-world situation where only a small number of labels are available in streaming data. In this paper, we propose a semi-supervised leaning approach that will augment training data samples and apply a fuzzy SVM algorithm. The augmented training technique automatically extracts and enlarges training set from the unlabelled streaming text, while learning is conducted by utilising a very small training set provided as an initial input. The experimental results indicate that the proposed augmented approach outperformed all other methods, and is suitable in the real-world situations, where sufficiently labelled instances are not available for training. For the proposed fuzzy SVM approach we handle complex and multidimensional data generated by streaming text, where the importance of features are discriminated for the decision function. The evaluation conducted on different experimental scenarios indicates the superiority of the proposed fuzzy SVM against all other methods.

Keywords: Cyberbullying Detection, Text-Stream Classification, Semi-supervised learning, Social Networks.

1 Introduction

Current studies on cyberbullying detection are mainly focused on: i) Supervised learning approaches, which rely on a human-intensive labelling process of data. ii) Feature space is uniformly applied to a learner. Whereas, streaming text generated by Social Networks (SNs) is highly uncertain, noisy, and imbalanced. In such a changing environment, different training data samples may have varying levels of importance. Therefore, with the rapid growth of user-generated content in SNs, existing supervised approaches become unaffordable and impractical for

---

* v.nahar@uq.edu.au

H. Wang and M.A. Sharaf (Eds.): ADC 2014, LNCS 8506, pp. 160–171, 2014.

automatic detection of cyberbullying instances. In this paper, we focus on the detection of cyberbullying in streaming text generated by SNs. For such detection the following challenges are identified:

(i) **Insufficient training instances:** Streaming text arriving from SNs, is either seldom labelled or not labelled at all. Moreover, it is impractical to ask users to label the messages into cyberbullying and non-cyberbullying categories. Therefore, as an alternative to manual labelling of the entire streaming text, only a small set of labelled instances are available for training.

(ii) **Uncertain and imbalance feature distribution:** All the input features is not evenly important for the learners' decision function. For example, the baseline swear-keyword based feature 'hell' is often used in normal communication. Using such words may not increase the discriminating effectiveness of the learner. To mimic the real-world situation, highly unstable and imbalanced data are fed to the system.

To address above challenges, we emphasis on cyberbullying detection under semi-supervised learning approaches. For this work, we assume that only a small set of labelled instances are available for initial system training. We consider two methods: i) based on augmented training by using ensemble classifiers with a confidence voting function. A confidence voting function is defined based on the parameter Γ to extract and enlarge training set from the unlabelled streaming text automatically. ii) using a fuzzy SVM algorithm to cater for the uncertain or irrelevant nature of the dynamic and multidimensional input feature space. In FSVM there are N possible free parameters (N), where N is the total number of training points. A degree of importance s_i is given to each data point providing a greater flexibility and generalisation to the model. For each training pair (x_i, y_i) a membership value s_i is given; the pairs with high s_i values will have a greater influence in the decision surface compared to the one with lower s_i values.

2   Related Work

Recently, Xu et al. explored regret behaviours in bullying messages assuming that people who posted bullying tweets may later want to delete those posts [1]. They reported cross validation accuracy upto 60.7%. Dadvar et al. used content-based, cyberbullying, and user-based feature sets [2]. The best recall obtained (55.0% recall, 77.0% precison, and 64.0% F_1 measure) with user-based and pronoun-profanity window feature sets. Dinakar et al. deconstructed cyberbullying detection into sensitive-topic detection, which is likely to result in bullying discussions, including sexuality, race, intelligence, and profanity[3]. Using SVM, the accuracy archived is 79% under the topic sexuality. Nahar et al. utilised probabilistic features and user ranking, and achieved 99% accuracy [4]. Yin et al. utilised various features including content, sentiment, and contextual features, showing 59.5% recall, 35.2% precision, and 44.4% F_1 measure [5].

However, these methods are conducted under supervised learning by directly applying the whole input feature space to a learner. These techniques are unable to handle the imbalanced and noisy data, where some features are either irrelevant or less important for the decision function. In this paper, we introduce semi-supervised learning for cyberbullying detection in streaming text.

3   Methodology

3.1   Feature Space Modeling

To understand the semantic structure users could have in mind while posting a comment, various features can be helpful. An enriched feature set can be generated from the given posts. These features are commonly known as linguistic features and are used mostly in nature language processing applications. These features are defined as follow: (i) Keywords based features, which involve binary representation of the keywords to see if the keywords are presented or not; (ii) To capture the influence of malevolence within messages, we also used the normalised value of the keywords. It is the number of swearwords in posts, divided by the total number of the words in messages; (iii) Presence of pronouns such as, 'you' and 'he' which makes the message more personal. For instance, if the keyword appears near 'you', it will likely indicate that the message is more targeted towards that person. Yei *et al.* used pronouns as sentiment features [5] for harassment detection, where normalised values of second and third person pronouns are used; (iv) To capture a degree of users' emotions, emotions are included for the feature space design. Normalised values of happy and angry emotions are computed separately for each comment; (v) Mostly, people on the Internet use capital letters to indicate that they are yelling or shouting. The normalised value of capital letters within messages is used to capture the loudness; (vi) Some other meta data of messages, such as special characters, are used in their normalised form; and (vii) Users' age and gender are also used as features because the selection of words, usage, and language vary between people of different age groups and gender.

In addition to above features, we also extracted location information. However in the dataset, most users come from different places of the USA, which would not carry ethnic or cultural differences since they are all from the same country. Therefore, location-based features are not considered for such datasets. However, it will be interesting to include location information to capture a certain degree of the ethnic or cultural differences and the language styles used.

3.2   Cyberbullying Detection

Following the traditional practices for performing text-stream classification [6][7], we assume that the unlabelled data streams U_n of varying length are arriving on the system in sessions. A very small set of positive instances and some negative instances are available for initial training $(P_n \cup N_n)$. To ensure that negative instances do not undermine the decision function for positive instances, random under-sampling of negative instances is adopted at the initial training phase. Every session is trained on two base classifiers. The model automatically extracts strong positive and some negative instances (T'_n) to enlarge the training set (T_n). The extraction and enlargement step is computed by combining the decision of both the classifiers using voting. To combine the decisions of multiple classifiers by voting, there are various ways such as linear combination, majority voting etc. that can be used. Out of those linear combination is the simplest way to combine multiple classifiers. According to the linear combination of the learners:

$$y_i = \sum_j \alpha_j d_{ji} \quad where, \quad \alpha_j \geq 0, \sum_j \alpha_j = 1 \tag{1}$$

Another possibility to find w_j is by assessing classifiers accuracy from separate feature set and use the information to compute the weight. We define *confidence function* Φ to predict the class label of the input instance given by Equation 2:

$$\Phi = \prod_j \alpha_j d_{ji} \quad where, \quad \alpha_j \geq 0, \sum_j \alpha_j = 1 \tag{2}$$

In the *confidence function* (Φ) α_j is the weight of the vote of the base classifier j for class C_i and d_{ji} is the vote of base classifier j for class C_i. Φ is defined based on the product of the probability distribution for the class C_i for the given test instance. As we are interested in cyberbullying posts, we select the predicted positive class based on the probability distribution of the base classifier j.

$$y_i = \begin{cases} 1 \ if \ \Phi \quad \geq \quad \Gamma \\ -1 \ if \ \Phi < (1 - \Gamma) \end{cases} \tag{3}$$

In U_n, cyberbullying instances are rare compared to normal or non-cyberbullying instances therefore, adding all the identified negative instances may results in overfitting. Thus, for each added positive instance $y_i = 1$, only two negative instances $y_i = -1$ will be augmented in the enlarged training set, T'_n.

We employ two base classifiers g_1 and g_2, which are well-known text classifiers. g_1 is a Naive Bayes multinomial text classifier, and the second base classifier g_2 is a Stochastic Gradient Descent classifier. Both classifiers are extended from WEKA[1], as they are available in WEKA under function-based algorithms.

The second base classifier g_2 is built using the Stochastic (or "on-line") Gradient Descent for text (SGD text) classifier [8] as an implementation in WEKA. The SGD text algorithm is designed for the text data. It generates feature space using an STWV filter to transform text strings into term-weight vectors based on Vector Space Model [9]. We use default values of the SGD text i.e. support vector machines as the loss function, learning rate of 0.01, a regularisation constant of 0.01, 500 iterations without pruning the dictionary, 3 as a minimum word frequency, default normalisation and no transformation to lower case of the input instances. The tokenisation of text strings is performed with the tokenise module specified by a parameter. It is an iterative method, which builds the learning model iteratively i.e. training is conducted in successive sessions until the algorithm converges using a loss function.

---

[1] http://www.cs.waikato.ac.nz/ml/weka/

Algorithm 1 : Model building using Augmented training set

Input:
  P_n: Small set of positive instances for initial training;
  N_n: Small set of negative instances for initial training;
  U_n: Set of unlabelled examples of the incoming session;
  Γ : *parameter*;
Output:
  T_n: Enlarged strong training set
1. $T_n \leftarrow P_n \cup N_n$;
2. //Extraction step:
3.   Train C_1 by T_n;
4.   Train C_2 by T_n;
5. $T_n' \leftarrow \{C_1, C_2\}$, Γ, using Equation 3;
6. //Enlargement step:
7. $T_n \leftarrow T_n \cup T_n'$;
8. *return* T_n;

By Using Fuzzy Approach: Given that the streaming text generated from the SNs is highly uncertain, complex, and unbalanced, we incorporate a membership generation method of the robust FSVM model. For the given array of text features for each user's comment, the method should have strong discriminatory power capable of ranking the input feature space. There are various methods used to generate membership values [10], [11], [12]. We use a Kernel-based Fuzzy C-Means (K-FCM) clustering algorithm to generate memberships values for our fuzzy classifier model owing to its ability to handle noise and outliers.

Clustering Process: The first step in K-FSVM is to cluster the incoming pre-processed text-stream data sets. In a complex and dynamic environment such as SNs, a range of features can be generated from single user post. Each feature will have a different degree of information and relevance to a specific concept therefore, calculating the total relevance of each instance from all features is highly important for the learning model. To achieve this goal we employ a fuzzy clustering approach, which enables us to evaluate all features and calculate their degree of relevance to a specific group. Clustering is used to find high intra-cluster and low inter-cluster similarities. The idea is to find natural groupings among similar objects.

Kernel-based FCM was introduced to overcome noise and outliers sensitivity found in FCM [13], [14] by transforming input space X to a high or infinite dimension feature F space ($\phi : X \rightarrow F$). For non-linearly separable problems, the input data can be projected to a high-dimensional feature space using a kernel. According to Cover's theorem, projecting input data into a high dimensional feature space is assumed to convert non-linearly separable problems into linearly separable in the feature space. This idea has been utilised in unsupervised learning and by many algorithms including RBF Networks, SVM and other non-linear discriminating techniques. In a Kernel FCM the input space is projected to higher dimensional feature space using RBF, polynomial kernel or any other kernel type.

A Kernel-based Fuzzy C-Means clustering (KFCM) algorithm has been proposed by Zhang et $al.$ [10][11]. KFCM partitions a given data set $X = \{x_i, ..., x_n\} \in R^p$ into $Cfuzzy$ fuzzy subsets by minimising the following objective function:

$$J_m(U,V) = \sum_{i=1}^{c} \sum_{k=1}^{n} u_{ik}^m ||\phi(X_K) - \phi(V_i)||^2 \qquad (4)$$

Subject to:

$$\sum_{k}^{n} u_{ik} > 0, \forall i \in 1, ...c \qquad (5)$$

$$\sum_{i}^{c} u_{ik} > 1, \forall k \in 1, ...n \qquad (6)$$

where, c is the number of clusters determined initially $(1 < c < n)$. According to the condition in Equation 4 no cluster is empty; n is the number of data points; u_{ik} is the membership of X_k in class i satisfying $\sum_{i}^{c} u_{ik} = 1$ for all k and $u_{ik} \in [0,1]$; m is the quantity controlling cluster fuzziness $(m > 1)$; V is a set of control cluster centres or prototypes $(V_i \in R^p)$; ϕ is an implicit nonlinear transformation function. The Euclidean distance between points and centres in the feature space F can be computed as:

$$||\phi(X_K) - \phi(V_i)||^2 = k(X_K, X_K) + k(V_i, V_i) - 2k(X_K, V_i) \qquad (7)$$

where, $K(X,Y) = \phi(X)^T \phi(y)$ is an inner product of the kernel function X denotes the data space, and $\phi(x) \in F$, where $x \in X$, F is the transformed feature space and $K(x,x) = 1$. In our case, a Guassian Kernel was adopted, where $K(x,y) = exp(-d(x,y)^2/^2)$. Hence for $K(x,x) = 1$, the Gaussian Kernel leads to $d\phi^2(x,y) = K(x,x) + K(y,y) - 2K(x,y) = 2(1 - K(x,y))$. Thus the objective functions in Equation 4 becomes:

$$J_m(U,V) = 2 \sum_{i=1}^{c} \sum_{k=1}^{n} u_{ik}^m (1 = k(X_K, V_i)) \qquad (8)$$

where,

$$k(X_K, V_i) = exp(-||X_K - V_i||^2/\sigma^2) \qquad (9)$$

The optimisation problem is solved by minimising $J_m(U,V)$ under the constraints of u_{ik}.

$$u_{ik} = \frac{(1/(1 - k(X_K, V_i)))^{1/(m-1)}}{\sum j = 1^c (1/(1 - k(X_K, V_i)))^{1/(m-1)}}, \qquad \forall i \in 1...c \ and \ \forall k \in 1...n \quad (10)$$

$$v_i = \frac{\sum_{k=1}^{n} u_{ik}^m K(X_K, V_i) X_K}{\sum k = 1^n u_{ik}^m K(X_K, V_i)} \qquad (11)$$

One of the critical steps in Fuzzy based SVM is the membership generator method. From the previous step the membership matrix is generated and used by a fuzzy classifier decision function. A good membership matrix should degrade the effect of outlier and noise, and improve overall classification results. The following algorithms are used to generate the membership matrix for the fuzzy classifier:

Algorithm 2: Kernel Fuzzy C-Mean Clustering Algorithm

Input:
b_n: Streaming text;
m: Set Fuzzification parameter;
c: Number of clusters;
ε_1:set termination parameter;
Output:
Membership matrix
1. Select the kernel function K and its parameters;
2. Select cluster centres v_i
3. Update membership matrix u_{ik} using Equation 10;
4. Compute all new clusters or prototype V_i using Equation 11;
5. Repeat step 3-4 and check the termination function E^t;
6. $E^t = max|v_n ew - V_o ld|$, if $E_t \leq \varepsilon$, **stop**;
7. *return* membership matrix;

Fuzzy Classifier: The enriched feature space generated in cyberbullying context will have some training points with a varying level of importance. Consequently, the training points with higher impact should be classified correctly and the noisy points or meaningless ones will not be considered and discarded. This indicates that one point can belong 85% to one class and the remaining 15% can be meaningless or 10% can belong to one class and 90% can be meaningless. In contrast to SVM, fuzzy SVM allows each data point X_i to be assigned a membership value, U_i where $0 < U_i \leq 1$. The membership U_i is used to determine the importance or relativity of each data point X_i to one class and the value $(1 - U_i)$ can be used to determine the degree of meaningless.

In this section, a fuzzy classifier is used to handle unbalanced and unstable text streams generated from social networks. The dataset is fed into the KFCM model to extract membership values, and then a one-versus-one (OVO) fuzzy SVM model is constructed. Each membership value is used by the FSVM for classification. It is expected that noise will be assigned a low membership degree and each membership value will be used by the FSVM model resulting in better generalisation and accuracy. The following steps are performed to execute the K-FSVM model:

Algorithm 3 : Fuzzy SVM Classification algorithm

Input:
b_n: Streaming text;
s_n: Membership matrix generated by Algorithm 2;
ε_1: Set termination parameter;
Output:
Final prediction matrix;
1. Use *OVO* strategy to create multiple classifiers initialize all parameters
   including kernel function, cluster number, termination parameter ε
   and membership m;
2. Apply the memberships s_n to FSVM model. Here, each data point will
   have one membership value.The new training set will have x, y, u,
   where u is the membership value for x data point.
3. Predict all class labels using voting by classifying which classes are
   receiving most voting;
4. *return* final classifier;

---

4 Experiments Setting

4.1 Dataset

In the experiments, we utilised data provided by Fundacion Barcelona Media[2]
for the workshop on content analysis from the Web 2.0. The given data was
collected from the three different SNs including *Myspace, Kongregate,* and
Slashdot. Characteristics of data from these three sites are different from each
other. Our task was to extract cyberbullying instances from the streaming text of
any type. The raw data was available in XML file format of different structures.
A data parser was developed to extract the content, time, and user information.
During the feature space modelling, extensive pre-processing was conducted in
order to remove insignificant features.

4.2 Evaluation

The classification of cyberbullying messages is a critical issue because of different
impacts made by the false positives and the false negatives. On one hand, to
identify non-cyberbullying instances as cyberbullying itself is a sensitive issue
(the false positive). On the other hand, the system should not miss out the
cyberbullying posts (the false negatives). Though the false positive and the false
negative instances are both critical, the ideal scenario is to achieve a high recall.
That means cyberbullying-like posts should not be overlooked by the system

---

[2] http://caw2.barcelonamedia.org/ Retrieved 10 November 2010.

- a strict approach. Nevertheless, we present a performance metric including precision, recall and F_1 *measure* for evaluation. Table 1 shows the distribution of positive instances, r in the training and testing dataset experimental setting for various scenarios used.

4.3 Results and Discussions

Experiment 1: We employ the session scenario by sorting messages using time information and generated N streaming sessions of varying length. For experiments, we select parameters, $N = 75$ and $\Gamma = 0.95$. The final training set constructed by the augmented training method is used as a training set. To evaluate this model on a test set (manually validated test set), we employ Random Forest, Naïve Bayes, Logistic Regression, and Meta classifiers, and results are shown in Figure 1(a). Expert judgement is also presented to compare the other classifiers. The majority of feature selection methods work better if the frequency of the positive-like features is high. From Figure 1(a), we observe that the model is able to capture likely positive words including words that appear in the keyword list. While detecting cyberbullying in social networks, recall is a critical evaluation matrix as it is very important to reduce false negatives. The system should not be able to misclassify positive cases as negative - that is, cyberbullying-like cases are not ignored. As shown in Figure 1(a), Random Forest performed similarly to that of the expert judgement, with 48% precision, 77% recall, and 59% F_1 *measure*. In this experiment, the false positive is higher than the false negative. If we try to reduce the false positive, then the false negative increases. This is because discrimination of the positive features and the negative features is very vague. In the training data, we observe that many likely cyberbullying words are quite frequent in both cyberbullying and non-cyberbullying categories. Ignoring those words on one hand reduces the false positives, while on the other hand it increases the false negatives. Our objective is to reduce the false negatives; therefore our system tolerates the false positives but maintains low false negatives. Nevertheless, in this experiment the objective was to achieve high Recall, which is achieved up to 79.3%.

Table 1. Positive Instance Distribution, r

Experiment	r	
	Training	Testing
Scenario 1	34.0%	13.2%
Scenario 2	42.1%	14.2%
Scenario 3	34.0%	1.5%
	10-fold cross validation	
Scenario 4	1.5%	
Scenario 5	22.0%	

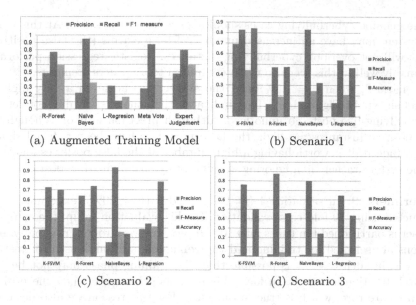

(a) Augmented Training Model   (b) Scenario 1

(c) Scenario 2   (d) Scenario 3

Fig. 1. Results of Experiments 1 and 2

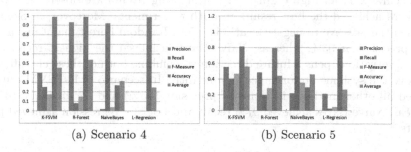

(a) Scenario 4   (b) Scenario 5

Fig. 2. Results of Experiment 3

Experiment 2: In this experimental set up, experiments are conducted in three different scenarios, when $r = 13.2\%$, 14.2%, and 1.5% respectively. The results are shown in Figures 1(b), 1(c), and 1(d).

In Scenario 1 ($r = 13.2\%$), K-FSVM outperformed all other methods. Recall is 82% while maintaining 6% precision. Nevertheless, Naïve Bayes achieved the highest recall (83%), which is almost the same as K-FSVM, whereas, Random Forest and Logistic Regression both performed similarly well.

In Scenario 2 ($r = 14.2\%$), the overall performance of K-FSVM and Random Forest are almost same, with the F_1 *measure* at 41%. While K-FSVM achieved higher recall, Random Forest achieved higher precision. Naïve Bayes achieved a 93% recall, which is one of the major requirements for a cyberbullying detection system. As discussed earlier, we are interested in high recall. That means the actual number of genuine cyberbullying cases identified by the system out of all the genuine cyberbullying cases should be as high as possible. At this stage, the system may have an increased number of false alarms (the false positive), but it will not overlook cyberbullying instances. Logistic regression performed reasonably well in this scenario when compared to all others scenarios.

In Scenario 3, when $r = 1.5\%$ only, the system has achieved a very high recall as shown in Figure 1(d). This indicates that the system did not let the cyberbullying-like posts go unnoticed, although precision is poor. Though the false positive instances are high, the false negatives have been reduced significantly. Indeed, it is worth having a high number of the false positives identified by the system rather than ignoring genuine or cyberbullying-like posts.

Experiment 3: In scenarios 4 (r=1.5%) and 5 ($r = 20\%$), experiments are conducted using 10-fold cross-validation. In this evaluation setup, the complete dataset is partitioned ten times into 10 samples. In every round, randomly, nine sections are selected for training and the remaining section is used for testing. However, in such cases it is possible that the training phase may not be able to catch positive instances. In fact, this likelihood increases when the positive instances are rare, which is true in our case. For this reason we also decide to compare overall performance, which is an average of precision, recall, F_1 *measure* and accuracy. From Figure 2(a), very interesting results are observed. Overall K-FSVM achieved the best results in both experiments. Moreover, in Scenario 4, when the positive to negative ratio is 1.5%, Random Forest maintains a very high precision at 93%, whereas, Naïve Bayes achieved the highest recall 92%. Such observation opens a future direction to combine both of these classifiers to improve the systems performance significantly. In scenario 5, K-FSVM outperformed all other methods in terms of precision (55%) and F_1 *measure* (47%), whereas, Naïve Bayes achieved 97% recall and Logistic regression achieved poor results.

5   Conclusions

This paper proposed a semi-supervised approach for detection of cyberbullying in SNs. Our contributions can be summarised as: (i) We devised a new framework for automatic detection of cyberbullying for the streaming data with insufficient labels. The framework extracts reliable positive and negative instances by augmented training methods based on the confidence voting function. (ii) The enriched feature sets were generated based on user context, linguistic knowledge, and baseline keywords were also incorporated during feature space design in the proposed method. (iii) We also proposed a fuzzy SVM algorithm for the effective cyberbullying detection. The proposed method effectively tackles the dynamic and complex nature of the streaming data. (iv) The experiments conducted under the different scenarios demonstrate that the proposed technique outperformed the traditional methods use for cyberbullying detection.

References

1. Xu, J.M., Burchfiel, B., Zhu, X., Bellmore, A.: An examination of regret in bullying tweets. In: The 2013 Conference of the North American Chapter of the Association for Computational Linguistics: Human Language Technologies, pp. 697–702 (2013)
2. Dadvar, M., Trieschnigg, D., Ordelman, R., de Jong, F.: Improving cyberbullying detection with user context. In: Serdyukov, P., Braslavski, P., Kuznetsov, S.O., Kamps, J., Rüger, S., Agichtein, E., Segalovich, I., Yilmaz, E. (eds.) ECIR 2013. LNCS, vol. 7814, pp. 693–696. Springer, Heidelberg (2013)
3. Dinakar, K., Reichart, R., Lieberman, H.: Modeling the detection of textual cyberbullying. In: AAAI Conference on Weblogs and Social Media, pp. 11–17 (2011)
4. Nahar, V., Unankard, S., Li, X., Pang, C.: Sentiment analysis for effective detection of cyber bullying. In: Sheng, Q.Z., Wang, G., Jensen, C.S., Xu, G. (eds.) APWeb 2012. LNCS, vol. 7235, pp. 767–774. Springer, Heidelberg (2012)
5. Yin, D., Xue, Z., Hong, L., Davisoni, B.D., Kontostathis, A., Edwards, L.: Detection of harassment on web 2.0. In: Content Analysis in the Web 2.0 Workshop at WWW (2009)
6. Zhang, Y., Li, X., Orlowska, M.: One-class classification of text streams with concept drift. In: ICDMW, pp. 116–125 (2008)
7. Nahar, V., Li, X., Pang, C., Zhang, Y.: Cyberbullying detection based on textstream classification. In: AusDM (2013) (in press)
8. Zhang, T.: Solving large scale linear prediction problems using stochastic gradient descent algorithms. In: ICML, pp. 919–926. ACM (2004)
9. Sebastiani, F.: Machine learning in automated text categorization. ACM Computing Surveys 34(1), 1–47 (2002)
10. Zhang, D.Q., Chen, S., Pan, Z.S., Tan, K.R.: Kernel-based fuzzy clustering incorporating spatial constraints for image segmentation. 4, 2189–2192 (2003)
11. Zhang, D.Q., Chen, S.C.: A novel kernelized fuzzy c-means algorithm with application in medical image segmentation. Artificial Intelligence in Medicine 32, 37–50 (2004)
12. Krishnapuram, R., Keller, J.M.: A possibilistic approach to clustering. IEEE Transaction on Fuzzy Systems 1, 98–110 (1993)
13. Wong, C.C., Chen, C.C., Yeh, S.L.: K-means-based fuzzy classifier design. 1, 48–52 (2000)
14. Gröll, L., Jäkel, J.: A new convergence proof of fuzzy c-means. 13, 717–720 (2007)

Mining the Association of Multiple Virtual Identities Based on Multi-Agent Interaction

Le Li[1,2], Weidong Xiao[1], Changhua Dai[1], Haiming Tong[2], and Zhiqiang Song[2]

[1] College of Information System and Management, National University of Defense Technology, Changsha, China
[2] China Satellite Maritime Tracking and Control Department, Jiangyin, China
lile10@126.com

Abstract. Abuses of online anonymity make identity tracing a critical problem in cybercrime investigation. To solve this problem, this paper focuses on the feature of authors' behavior in time slices and tries to mine the association of multiple virtual identities based on multi-agent interaction. We propose the recognition model MVIA-K based on knowledge management. In MVIA-K, agents perform distributed mining to get candidate author groups as local knowledge in each time slice. Then high-quality knowledge is extracted from the local knowledge and used as priori knowledge to guide other agents' mining process. Finally distributed knowledge is integrated on the basis of knowledge scale. Experiment with real-world dataset shows that MVIA-K has a very promising performance, which can filter the noise data effectively and outperform Author Topic model.

Keywords: Virtual identities, time slice, multi-agent, text mining.

1 Introduction

The unprecedented development of the Internet makes forum, blog and microblog get widespread attention and become the main platforms for people to communicate. However, due to the online anonymity, it is not uncommon for people to maintain numerous virtual identities [1]. For example, some famous people post articles in several websites by using different accounts in order to expand their influence. Some people take use of the online anonymity to achieve their illegal purposes, such as mail fraud [2], improving reputation rank by creating fake sales [8]and so on. Spammers tend to use multiple virtual identities to publish numerous articles in order to influence search engine results or affect public opinion [7,4]. Therefore, it is valuable to recognize the association of multiple virtual identities.

Researchers have begun to use textual traces to find the association of virtual identities [11,1]. These methods commonly use the stable feature of the author - writing style, but they do not concern the feature of user's behavior. This kind of authors has an obvious behavior feature in time slice, who tend to use multiple virtual identities to post similar articles in multiple time slices.

H. Wang and M.A. Sharaf (Eds.): ADC 2014, LNCS 8506, pp. 172–179, 2014.

In this paper, we believe that time slice plays an important role in recognizing the association of these virtual identities.

1. By analysis on time slice, we can find the latent association of these virtual identities. In order to avoid inspection, spammers, who own numerous virtual identities, will use different identities to accomplish different tasks [4]. On the whole these virtual identities have no obviously association, but from the author's behavior feature, these virtual identities tend to post similar articles in the same time slice. Only with time slice analysis can we recognize association of virtual identities.

2. By introducing time slice, we can filter the noise data effectively. Hot topic will attract many authors in the discussion in a period. Although these virtual identities post similar articles to the same topic, they do not actually belong to the same author. When time slice is introduced, we can find that the noise data (unrelated IDs) will be only associated with target identities in just one slice, but target identities belong to the same author will have strong association in multiple time slices.

This paper tries to mine the association of multiple virtual identities based on multi-agent interaction. We design a recognition model MVIA-K based on knowledge management. Agents perform distributed mining to find the candidate author groups in each time slice, and then high-quality knowledge will be extracted and transferred to other agents. Agents take use of it as priori knowledge to guide the subsequent mining process, which can effectively remove the impact of noise data. Finally we integrated distributed mining results based on knowledge scale.

2   Related Works

In this paper, we use topic analysis method to compare the similarity of virtual identities. LDA [3] is a widely used topic model, which can model documents to word-topic distribution and topic-document distribution. Author Topic model [12,15] is a probabilistic topic model to study the relationship between author and text ,but it did not consider the authors with time information. Paper [6] presents Temporal-Author-Topic (TAT) approach to model author's interests and time of documents. However, we need to focus on users' behavior feature in time slice, but TAT cannot capture it. Therefore, we do not create a probabilistic model, but integrate the agent and data mining technology to build a recognition model based on multi-agent interaction.

Authorship attribution tries to assign a text of unknown authorship to one candidate author [13] , paper [9] concerned the potential textual traces of identity. Paper [1] developed the writeprints technique for similarity detection of anonymous identities. These researches use writing style to identify the author. But for spammer detection, organizers will employ good writers to prepare specific post templates for spammer [4]. Spammers just need to make appropriate modifications, which weakens the difference of writing styles. So these methods are not suitable for recognizing the association of these virtual identities.

Paper [10] tried to recognize the association of virtual identities by using the time slice. However, in their method, only part of the local results are reported and integrated, which may miss some valuable local knowledge; secondly, they integrated local results by voting, but did not consider the confidence of different results.

3   Recognition of Multiple Virtual Identities Association

In this section, we give a specific description about the problem of recognizing multiple virtual identities association, and indicate the difficulties. Then we propose: MVIA-K, a recognition model for the association of multiple virtual identities.

3.1   Problem Formulation

This kind of users in cyberspace usually have multiple virtual identities and they tend to choose different identities for posting in different periods of time.

We assume the virtual identities set of a user is: $U_{all} = \{ID_1, ID_2, \ldots, ID_n\}$. When the user needs to post in T_i, he will randomly choose a subset $U_i = \{ID_k, ID_l, \ldots, ID_o\} \subseteq U_{all}$ for posting. Each ID in U_{all} publishes articles of different topics with great randomness and we cannot find any obvious association. However, by introducing time slice, we can find that IDs in U_i have a strong relationship in the time slice T_i, make it relatively easy to recognize. So our goal is to look for the subset U_i in time slice T_i, and then integrated to get U_{all}. But some hot topics will attract a lot of people (including our target authors) to participate in the discussions, which led these IDs publish similar content over a period of time. It brings difficulty for us to recognize because of more noise data (unrelated IDs). So sometimes we get U_i' with noise data rather than target U_i in T_i. How to use U_i' to get U_{all} is the key point of this paper.

3.2   The Choo Sense-Making KM Model

In this paper, we use the idea of the Choo Sense-making KM Model [5] for recognition of multiple virtual identities and design MVIA-K model from the perspective of knowledge management.

Choo has described a model of knowledge management that stresses sense making, knowledge creation and decision making. In the sense-making stage, one attempts to make sense of the information streaming in from the external environment. Knowledge creating may be viewed as the transformation of personal knowledge between individuals through dialogue, discourse, sharing, and storytelling. Decision making is situated in rational decision-making models that are used to identify and evaluate alternatives by processing the information and knowledge collected to date.

3.3   Recognition Model: MVIA-K

We have introduced the challenges in the above section. In this section, we proposed MVIA-K, a recognition model based on multi-agent interaction. Guided by the theory of knowledge management, we use knowledge extraction, flow and integration process to achieve effectively recognition purposes. The model is shown in Fig.1.

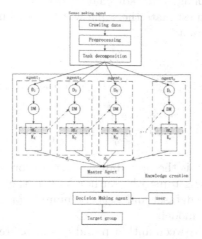

Fig. 1. Recognition model

In MVIA-K, first of all, sense-making agent actively perceives the dynamic needs of users and receives user-specified tasks. According to task, agent captures relevant data and realizes data preprocessing. Then sense-making agent decomposes the tasks according to the specific time slice size and informs other agents to start subsequent knowledge mining.

Knowledge creating agents perform distributed data mining in their own dataset D_i to obtain the local knowledge K_i. Then high-quality knowledge HK_i can be extracted from K_i and passed to other agents. Agents received HK_i will use it as priori knowledge to guide their subsequent mining processes, which reduce the impact of the noise data. Master agent integrates distributed mining results based on knowledge scale d_i to get target group.

Decision making agent receives the results of mining process and forms group knowledge. Then agent can extract the target author group through human-computer interaction or user-defined thresholds.

4   Knowledge Based Multi-Agent Mining and Interaction

How to extract high-quality knowledge, how to make high-quality knowledge flow to guide subsequent mining process, and how to integrate local knowledge effectively, these are important factors affect the recognition ability of model. In this section, we will introduce the specific implementation process.

4.1   Local Knowledge Extraction

First of all, we need to find the association of authors and topic of their articles. AT model (Author Topic model) is an probabilistic model of Author-Topic. By analyzing the contents, author is mapping to a probability distribution of topic in AT model. Different topic distribution represents the author's knowledge structure and writing tendencies.

By using Gibbs sampling, a Markov chain Monte Carlo algorithm to calculate the distribution, we can estimate the topic-words distribution and author-topic distribution using equation:

$$\phi_{mj} = \frac{C_{mj}^{WT} + \beta}{\Sigma_{m'} C_{m'j}^{WT} + V\beta} \tag{1}$$

$$\theta_{kj} = \frac{C_{kj}^{AT} + \alpha}{\Sigma_{j'} C_{kj'}^{AT} + T\alpha} \tag{2}$$

In the equation, C_{mj}^{WT} is the probability of using word m in topic j, and C_{kj}^{AT} is the probability of using topic j by author k. V is the size of the words, T is the topic number in the dataset and A is the number of authors. α and β is the hyperparameters of the model.

After that, distance between author p and q is measured by the symmetrized Kullback Liebler (KL) distance between topic distributions. As in equation 3, smaller distance value means higher correlation between the authors.

$$dis(p,q) = \frac{1}{2}[\sum_{j=1}^{T} p_j \log_2 \frac{p_j}{q_j} + \sum_{j=1}^{T} q_j \log_2 \frac{q_j}{p_j}] \tag{3}$$

We get local knowledge K={$A_1,A_2,\ldots,A_n$} by calculating KL distance of the authors. Then high-quality knowledge HK⊆K was extracted by setting a threshold value d, where ∀A∈ HK→dis(k,l)<d. In order to improve the influence of high-quality knowledge, we punish the virtual identities not in HK (adding a penalty value to its KL distance). After that, agent reports K to the master agent and passes HK to other agents.

4.2   Knowledge Flow

In the transfer stage, we use unidirectional transformation for simplicity, where each agent transfers the knowledge to only one agent and each agent receives priori knowledge only once. Since each agent's individual knowledge is different, in learning process, agent needs to check the priori knowledge priori-K to find valuable knowledge priori-K', and then use it to revise the preliminary mining results (we take the mean value of both distance). Knowledge flow in the model can expand the influence of high-quality knowledge and contribute to reducing the influence of local noise data.

4.3   Knowledge Integration

After local mining process, agents will negotiate with the master agent initiatively and cast their vote, which marked their local knowledge. Although each agent votes only once, it does not have the same weight on the final result. Obviously, compared to small dataset, the knowledge mining from large dataset should be more credible. In this article, we treat the amount of documents within each time slice as the knowledge scale. Master agent uses Equation 4 to calculate the distance between virtual identity k and l, where D_i is the amount of documents within the time slice T_i, $dis_i(k, l)$ is the distance between virtual identity k and l within the time slice T_i.

$$dis(k, l) = \sum_{i=1}^{n} \frac{D_i}{\sum_{i=1}^{n} D_i} dis_i(k, l) \tag{4}$$

5   Experiments

We selected 33 famous authors from three Chinese blog website (sina, souhu, 163) and collected 12,865 articles from 2009 to 2010. There are 12 target authors among them have more than one account in different website and post similar articles in time slice, which accords with definition of our problem. Meanwhile, these famous people have real-name authentication on these websites, so it will be easy for us to verify the experiment results.

We choose the AT model as benchmark technique and use the same hyper-parameters as [14] in the AT model, hyper-parameters α and β were set at 50/T and 0.01, the number of topics T=20, the number of iterations is 500.

In MVIA-K, KL threshold is used to extract high-quality knowledge and select the final target group, so we choose different KL threshold (from 0.3 to 0.7) in the experiment. Meanwhile, time slice size also exerts an influence on the results. If the size is too small, there will be only a few articles in same time slice, we cannot find any association of the virtual identities. But if the size is too large, we may lose user's behavior feature in time slice. In this paper, the dataset is divided into 5 and 10 time slices respectively, named MVIA-K5 and MVIA-K10. We use precision and F1 value as criterion in the experiment. The results are shown from Fig.2:

As can be seen, MVIA-K is significantly better than the AT model in a number of KL thresholds. This is because some hot topics in cyberspace bring a lot of noise data, making many authors of the similar distribution in topic. AT model can filter noise data coarsely only relying on KL threshold (this is the reason of its high recall). KL threshold plays an important role in the recognition ability. Compared to AT model, our method can filter the noise data more effectively. But if the KL threshold is too large, the local results will be of low correlation, leading to the decrease of precision and F1 values.

We can find that different size of time slice has a certain impact on the results. MVIA-K10 performs better than MVIA-K5 in this dataset. This is because

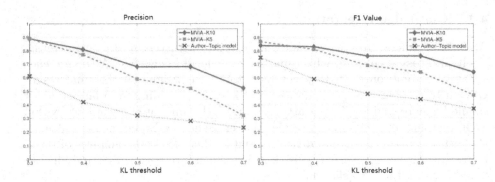

Fig. 2. Experiment Results

high-quality knowledge transfers more times in MVIA-K10, which ensures that poor local results have little effect. However, in MVIA-K5, poor local results will have a greater impact. So we can find that, when the KL threshold increases, downward trend of MVIA-K5 is more obvious. But it does not mean that setting a larger time slice size is always better. When the time slice division is excessive, the amount of articles within each time slice will be too small to extract high quality knowledge, which will make it difficult to get the valuable results. Therefore, in practical application, it will be more appropriate to choose multiple time slice size and get the mean result.

6   Conclusion

This paper focuses on the recognition of multiple virtual identities association. By using time slice to analyze user's behavior, we propose a recognition model based on knowledge management theory: MVIA-K. By integrating agent technology and data mining, the distributed knowledge can flow effectively in the model. High-quality knowledge can guide the subsequent mining process to obtain higher quality results. Meanwhile, knowledge scale based integration method can overcome the impact of noise data effectively and achieve better experimental results.

Acknowledgments. This work was funded under National Science and Technology Support Program (NO.2012BAH08B01) and the National Natural Science Foundation of China (No.61302144).

References

1. Abbasi, A., Chen, H.: Writeprints: A stylometric approach to identity-level identification and similarity detection in cyberspace. ACM Transactions on Information Systems (TOIS) 26(2), 7 (2008)

2. Airoldi, E., Malin, B.: Data mining challenges for electronic safety: The case of fraudulent intent detection in e-mails. In: Proceedings of the Workshop on Privacy and Security Aspects of Data Mining, pp. 57–66 (2004)
3. Blei, D.M., Ng, A.Y., Jordan, M.I.: Latent dirichlet allocation. The Journal of Machine Learning Research 3, 993–1022 (2003)
4. Chen, C., Wu, K., Srinivasan, V., Zhang, X.: Battling the internet water army: Detection of hidden paid posters. arXiv preprint arXiv:1111.4297 (2011)
5. Chun, W.C.: The knowing organization: How organizations use information to construct meaning, create knowledge, and make decisions. Oxford University Press, Oxford (1998)
6. Daud, A., Li, J., Zhou, L., Muhammad, F.: Exploiting temporal authors interests via temporal-author-topic modeling. In: Huang, R., Yang, Q., Pei, J., Gama, J., Meng, X., Li, X. (eds.) ADMA 2009. LNCS (LNAI), vol. 5678, pp. 435–443. Springer, Heidelberg (2009)
7. Gao, H., Hu, J., Wilson, C., Li, Z., Chen, Y., Zhao, B.Y.: Detecting and characterizing social spam campaigns. In: Proceedings of the 10th ACM SIGCOMM Conference on Internet Measurement, pp. 35–47. ACM (2010)
8. Josang, A., Ismail, R., Boyd, C.: A survey of trust and reputation systems for online service provision. Decision Support Systems 43(2), 618–644 (2007)
9. Li, J., Zheng, R., Chen, H.: From fingerprint to writeprint. Communications of the ACM 49(4), 76–82 (2006)
10. Li, L., Xiao, W., Dai, C., Xu, J., Ge, B.: The recognition of multiple virtual identities association based on multi-agent system. In: Cao, L., Zeng, Y., Symeonidis, A.L., Gorodetsky, V., Müller, J.P., Yu, P.S. (eds.) ADMI 2013. LNCS(LNAI), vol. 8316, pp. 40–50. Springer, Heidelberg (2014)
11. Pan, Y.: Id identification in online communities. Technical report. Citeseer (2006)
12. Rosen-Zvi, M., Griffiths, T., Steyvers, M., Smyth, P.: The author-topic model for authors and documents. In: Proceedings of the 20th Conference on Uncertainty in Artificial Intelligence, pp. 487–494. AUAI Press (2004)
13. Stamatatos, E.: A survey of modern authorship attribution methods. Journal of the American Society for Information Science and Technology 60(3), 538–556 (2009)
14. Steyvers, M., Griffiths, T.: Probabilistic topic models. Handbook of latent Semantic Analysis 427(7), 424–440 (2007)
15. Steyvers, M., Smyth, P., Rosen-Zvi, M., Griffiths, T.: Probabilistic author-topic models for information discovery. In: Proceedings of the Tenth ACM SIGKDD International Conference on Knowledge Discovery and Data Mining, pp. 306–315. ACM (2004)

Split Dictionaries for In-memory Column Stores in Mixed Workload Environments

David Schwalb, Markus Dreseler, Martin Faust,
Johannes Wust, and Hasso Plattner

Hasso Plattner Institute, Potsdam, Germany

Abstract. Columnar in-memory databases use dictionary encoding as
a compression technique, replacing long and frequently occurring values
with short integers. Sorted dictionaries allow for more efficient query
processing as comparisons can be performed directly on the compressed
data whereas unsorted dictionaries are faster when inserting new values.

In this work, we propose a new type of dictionary compression called
Split Dictionaries. These organize their values in fixed-sized splits, enabling fast inserts and comparable query performance while significantly
reducing maintenance costs. We present a detailed performance analysis
regarding inserts, range queries, and the merge process as well as a memory usage model. We argue that adjusting the dictionary size allows for
a more balanced trade-off especially in mixed workload environments.

1 Introduction

Traditionally, column stores are strong in analytical scenarios where workloads
are read-only and inserts are processed in batches. Recent work proposes their
use also for mixed workloads in enterprise environments [2, 4, 8, 9, 11, 12]. A
common approach is to use a read-optimized main and a write-optimized delta
partition which are periodically combined by a merge process [5, 13].

Both use dictionary encoding to replace long values with short integer codes.
An essential difference in the organization of main and delta partitions is the
order of assigned value ids. Main partitions use dictionaries sorted by value,
whereas delta dictionaries are unsorted [5]. Sorted dictionaries allow for fast
searches as values can be compared without decompression, but are unsuitable
for single insertions due to the sort order. In contrast, unsorted dictionaries need
to decompress the encoded values for comparisons but allow for fast insertions
as no sort order has to be kept. While main/delta architectures alleviate the
mentioned tradeoff, they increase the system complexity and require background
merge processes which use system resources [12]. The required rewrites, not the
update of the dictionary itself, are the major problem of sorted dictionaries.

In this paper, we propose Split Dictionaries as an alternative. Instead of using
a single dictionary, Split Dictionaries consist of a number of fixed-sized, sorted
dictionaries and a single, unsorted dictionary. With this architecture, Split Dictionaries provide a range query performance that is significantly better than that
of unsorted dictionaries. Additionally, the merge now only rewrites value ids that

H. Wang and M.A. Sharaf (Eds.): ADC 2014, LNCS 8506, pp. 180–188, 2014.

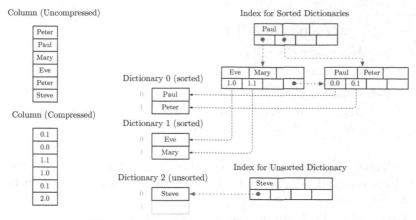

Fig. 1. Layout of data in a Split Dictionary

were inserted since the last merge, reducing the merge costs. We argue that Split Dictionaries allow for better balanced trade-offs especially for mixed workloads, resulting in increased overall performance. Our contributions are i) the concept of Split Dictionaries, ii) a memory usage model, and iii) a detailed performance evaluation comparing sorted, unsorted, and Split Dictionaries.

2   Split Dictionaries

This section describes Split Dictionaries and presents a memory model comparing them with sorted and unsorted dictionaries. It focuses on the implementation of range queries and on how the merge benefits from Split Dictionaries.

These are designed to combine the advantages of sorted and unsorted dictionaries. The basic idea is the separation into multiple dictionaries. For efficient insertions, new values are inserted into an unsorted dictionary. Once this dictionary reaches its maximum size, it gets sorted and is then immutable for the future. The maximum dictionary size is a power of two, thus mapping the dictionary id and local value id to a single value id can use simple bit operations.

To support efficient lookups, a tree is maintained that indexes the values in all sorted dictionaries. If a value is not found in this index, it is searched for in the unsorted dictionary, which is a tree in itself. By updating the main index when a dictionary is sorted, we reduce the overhead of the insert.

Figure 1 shows a column using Split Dictionaries after all insertions. We start with an empty column and a maximum dictionary size of 2. For simplicity, a B+-tree is shown instead of an rb-tree. We start with one dictionary with id 0. 'Peter' is inserted with the value id 0.0, next 'Paul' gets 0.1. As dictionary 0 is full, it is sorted and the value ids are changed in the attribute vector by replacing old value ids with new ones in the attribute vector, similar to the merge described in [5]. Also, the index on the sorted dictionaries is updated. 'Paul' now has the value id 0.0. Next, we create a new unsorted dictionary, insert the new values 'Mary'

$$M_A = (n \cdot \lceil \log_2 d \rceil)/8 \tag{1}$$
$$M_T(x) = x \cdot (s + 36) \tag{2}$$
$$M_{Sorted} = M_A + d \cdot s \tag{3}$$
$$M_{Unsorted} = M_A + d \cdot s + M_T(d) \tag{4}$$
$$M_{Split} = M_A + s \cdot (d - u) + u \cdot s + M_T(u) + M_T(d - u)$$
$$= M_A + d \cdot s + M_T(d) \tag{5}$$

and 'Eve', and merge it, resulting in the value ids 1.0 and 1.1. Finally, 'Steve' is inserted and assigned value id 2.0 in the newly created unsorted dictionary.

2.1   Memory Usage of Split Dictionaries

Besides performance, a classic trade-off for data structures is their memory usage. This section outlines the memory usage of Split Dictionaries. Let s denote the size (in Bytes) of an uncompressed value, n the number of entries in the column, d the number of distinct values, m the maximum dictionary size, and $u = n$ (mod m) the size of the unsorted dictionary in Split Dictionaries.

Equation 1 gives the size of a bit-packed attribute vector [14], and Equation 2 the size of an RB-Tree with x entries. The sorted dictionary (Eq. 3) is simple vector storing each distinct value. The unsorted dictionary (Eq. 4) additionally stores a tree on top of the unsorted vector, using additional memory. For Split Dictionaries (Eq. 5), $d - u$ values are stored in a sorted vector and u values in the unsorted vector. Additionally, we have the index on the unsorted dictionary with u values and the index on the sorted dictionaries with $d - u$ values.

As we can see, Split Dictionaries require more memory than sorted dictionaries due to the rb-tree and have the same memory costs as unsorted dictionaries.

2.2   Full Column Scans with Range Selections

Before we discuss range queries on Split Dictionaries, we will explain them for sorted and unsorted dictionaries [11]. Range queries on sorted dictionaries exploit the order of the value ids. A range of values translates to a range of value ids. By comparing value ids from the attribute vector with the value ids of the range boundaries, range queries can be performed with only two integer comparisons.

The naïve approach for unsorted dictionaries decompresses the column and compares the actual values - a very expensive step. Alternatively, a bit-vector can be used: First, for every value in the dictionary, the corresponding position is set to 1 if the value is in the range. Second, when iterating over the attribute vector, positions are included if the corresponding position in the bitvector is 1.

We will now describe the algorithm used for Split Dictionaries. Similar to the algorithm for sorted dictionaries, we iterate over all sorted dictionaries and find the bounding value ids. We end up with a list containing one tuple (min, max) for every sorted dictionary. If we had two dictionaries, the result could be $[(17, 25), (10, 13)]$. For every value id in the attribute vector, we extract the dictionary id and get the corresponding tuple from our auxiliary structure. We then check if the local value id is in the specified range. For value id 1036, the dictionary id is 1 and the local value id 12. We then get the tuple with the index 1, which is $(10, 13)$. Since 12 is in this range, the value id 1036 belongs to a value in the queried range. Checks are done in constant time as tuples can be accessed using pointer arithmetics. For value ids from the unsorted dictionary, however, we cannot use this approach and use the bit-vector algorithm instead.

2.3 Implications on the Merge Process

When a dictionary becomes full, we sort it and rewrite its value ids in the attribute vector. This is comparable to the merge for a column with main and delta partitions [5]. Assuming insert-only semantics (e.g., when MVCC is used), we reduce the number of attributes that have to be probed during the merge: The value id of a value changes only once. We thus only need to look at value ids in the vector that were inserted after the dictionary was created.

We also reduce the number of merges: With the current delta approach, a merge is needed whenever the attribute vector reaches a certain size. Since with Split Dictionaries we can append to the main attribute vector without needing a delta attribute vector, merges are only required if the unsorted dictionary is full. This is an advantage when dictionaries become saturated.

3 Evaluation

Our implementation uses libstdc++ vectors for the attribute vector and maps for the dictionaries. We used GCC 4.8 with -O3. Benchmarks were executed on a machine with 4 Intel Xeon 7560 CPUs (2.26 GHz) and 1 TB of main memory.

3.1 Insert Performance

To measure the insert performance, we inserted 100 million values with a uniform distribution of 10 million distinct values. The maximum dictionary size was 1024. We measured an overhead of 13% compared to unsorted dictionaries due to the costs of merging the unsorted dictionary and maintaining the global index.

3.2 Decompression of Value ids

Decompressing a value id is a pointer calculation for all dictionaries, adding the local value id to the beginning of the vector. Split Dictionaries need to separate the dictionary id from the local value id. This results in an overhead of 10% on a dictionary with 100 to 10k values. For bigger dictionaries, the cost is dominated by the memory reads, and the overhead gets as low as 1% for 100k entries.

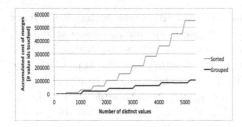 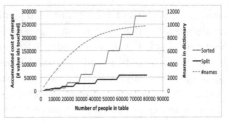

Fig. 2. Cost of the merge for (a) a linearly growing dictionary and (b) with a dictionary growing as modeled by a logistic function

3.3 Merge

We will now compare the merge costs of different dictionaries. In a first example, we take a constantly growing number of distinct values. For sorted dictionaries, we merge after 10k entries; for Split Dictionaries, we use a maximum dictionary size of 1024. For different configurations, the overall results are similar.

Figure 2(a) shows that Split Dictionaries need fewer rewrites for any table size. This, of course, is partially influenced by the merge criterion and the maximum dictionary size. Much more important, though, is that the cost of every single merge increases linearly with unsorted dictionaries, but remains constant with Split Dictionaries. In fact, choosing a lower maximum dictionary size would only add more "steps" to the graph of the Split Dictionary merge, but would not increase the accumulated cost of the merge. This is because for every merge, the existing approach looks at all existing entries, while the Split Dictionary can limit the search to the ones that have not yet been merged.

In a second evaluation shown in Figure 2(b), we show the merge costs when the dictionary is getting saturated. We store 10,000 different last names in a database. First, most entries result in a new name being added to the dictionary. Later, most names have already occurred. We model this using a logistic curve (represented by the dashed line) with $n(x) = \lceil \frac{1}{1+e^{-x/18000}} * 20000 - 10000 \rceil$ giving the number of distinct names after x inserts. We see a similar result with merges being more expensive and overall merge costs higher for sorted dictionaries.

3.4 Range Query Performance

The next experiment, shown in Figure 3, compares the range query performance. The distinctivity of the uniformly distributed values was 10% and the query selectivity 1% of the distinct values. We used a maximum dictionary size of 256.

Sorted dictionaries show an almost constant performance. Split Dictionaries have some overhead for small columns (1,000 entries) due to the additional cost of initializing the different parts of the range query. After this, their performance is also constant for up to 100,000 entries. Towards the end, they become slightly more expensive due to the increasing cost of building the interval vector. Once the attribute vector becomes larger than one million entries, Split Dictionaries

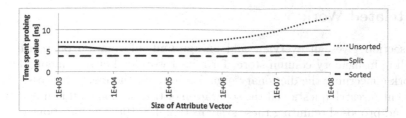

Fig. 3. Cost of Range Queries

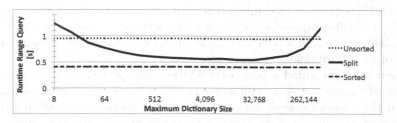

Fig. 4. Influence of the Maximum Dictionary Size for range queries on a table with 100M rows, 20% distinct values and 1% selectivity

have a significant and growing advantage over unsorted dictionaries. This is because for more than 262,144 distinct values, the bitvector for the unsorted dictionary requires 32 KiB, exceeding the L1 cache. The auxiliary structure needed for the Split Dictionary only requires 2 KiB and fits into the L1 cache.

While the algorithmic complexity is the same for all dictionaries, we get significant differences as a result of the number of steps and the cache behavior. Split Dictionaries perform better for range queries than unsorted dictionaries in any case, but especially for large columns. Their performance is, however, influenced by the choice of the maximum dictionary size.

3.5 Influence of the Maximum Dictionary Size

We will now evaluate the influence of the maximum dictionary size on the performance of range queries. The benchmark had 20% distinct values and a selectivity of 1%. The size of the attribute vector was 100M.

Figure 4 shows another trade-off, this time between small and large dictionaries. If the dictionaries are too small, the auxiliary structure becomes too large as it has to hold many tuples. In this case, it does not fit into the L1 cache anymore, resulting in many cache misses. If the maximum dictionary size is chosen too high, too many entries point to the unsorted dictionary and require the significantly slower bit-vector algorithm.

4   Related Work

This section gives an overview and background information on related work regarding in-memory column stores for mixed workload environments, followed by work concerning the dictionary organization and the merge process.

Recent research questioned the separation of transactional and analytical systems and proposed reuniting these systems [1,2,4,8–10]. The backbone of such a system is proposed to be a compressed in-memory column-store [1,4,9]. Column-oriented databases have proven to be advantageous for read-intensive scenarios [7,13,15], especially in combination with an in-memory architecture. Typically, lightweight compression techniques like dictionary encoding are used [6] as the challenge is to find a balanced trade-off between the contradicting requirements of mixed workload environments.

Recent work presented a parameter analysis and performance discussion for equality and range scans, as well as for positional lookups, inserts, and merges [11]. In this, uncompressed, bit-compressed, and dictionary-encoded columns with sorted and unsorted dictionaries were considered. The described merge algorithm depends linearly on the complete table size [5], whereas Split Dictionaries as proposed in this paper allow for significant cost savings regarding the merge process by only re-encoding values that were inserted since the last merge.

Sikka et al. see a major drawback of the merge in the CPU and memory costs caused by creating a new main [12]. They propose a partial merge strategy, splitting the main into independent main structures and only partially merging these. In contrast to our approach, from time to time a full merge is still necessary.

When dictionaries are shared between columns, changing sorted structures requires a re-encoding of multiple columns. Hildenbrand proposes a dictionary encoding scheme called Multi-Version Encoding, supporting insertions into a sorted dictionary by versioning the dictionary and providing a mapping between the different versions [3]. In contrast to our approach, periodic re-encodings of the complete column are still necessary as the mapping structures add a growing overhead for query processing. Especially for the use-case of global dictionaries, Split Dictionaries promise to be an attractive alternative.

5   Future Work

In addition to the advantages before, there are more concepts in which Split Dictionaries can improve in-memory databases. As of now, these have been evaluated theoretically. A thorough evaluation shall be subject to future work.

The current merge approach creates a new attribute vector into which the values from both main and delta are inserted. This is problematic for large tables, as the required memory doubles during merge processes. We argue that this is not necessary in Split Dictionaries. A modified sorting step allows for asynchronous, lock-free, and in-place sorting of the unsorted dictionary. The concept behind this is to create a new dictionary with a new dictionary id instead of invalidating the old dictionary. value ids in the attribute vector can be asynchronously changed from the old value id to the new one. Meanwhile, we can already insert new values in the new unordered dictionary. This approach requires a change in basically all operators that operate on the attribute vector. These operators have to be aware that there might be two value ids pointing to the same value.

Our experiments further suggest that Split Dictionaries can be helpful when using shared dictionaries to speed up the join process by sharing dictionaries between tables. When equal values have equal value ids, creating a mapping between value ids from one table to the other becomes obsolete. However, this can come with a high price if sorted dictionaries are used: If a new distinct value is inserted in either of the tables, both tables have to be rewritten. Since Split Dictionaries reduce the merge cost significantly, they can help to make Shared Dictionaries a more feasible approach.

6 Conclusions

In this paper, we proposed Split Dictionaries as a dictionary compression for in-memory column stores. These split the dictionary into a number of fixed-sized splits. Dictionaries are sorted once as they reach their capacity.

We presented a detailed performance discussion for Split Dictionaries, focusing on the performance of range queries, decompression, and merge costs. Split Dictionaries have a range query and lookup performance comparable to sorted dictionaries. Merge costs can be reduced significantly compared to the traditional approach which uses a sorted main and unsorted delta partitions. The cost of this is a small overhead for inserts and when decompressing values. Furthermore, we presented a memory usage model that shows that Split Dictionaries as much memory as unsorted dictionaries and more than sorted dictionaries.

In summary, we argue that Split Dictionaries allow to finely adjust the trade-offs in mixed workload environments, resulting in better overall performance and reduced merge costs. For future work we look into auto-adjusting the maximum dictionary size and in-place merge algorithms.

References

1. Färber, F., Cha, S.K., Primsch, J., Bornhövd, C., Sigg, S., Lehner, W.: SAP HANA database: Data management for modern business applications. SIGMOD (2012)
2. Grund, M., Krueger, J., Plattner, H., Zeier, A., Cudre-Mauroux, P., Madden, S.: HYRISE—A Main Memory Hybrid Storage Engine. In: VLDB (2010)
3. Hildenbrand, S.: Scaling Out Column Stores: Data, Queries, and Transactions. PhD thesis, ETH Zurich (2012)
4. Kemper, A., Neumann, T.: HyPer: A hybrid OLTP&OLAP main memory database system based on virtual memory snapshots. In: ICDE (2011)
5. Krüger, J., Kim, C., Grund, M., Satish, N., Schwalb, D., Chhugani, J., Plattner, H., Dubey, P., Zeier, A.: Fast Updates on Read-Optimized Databases Using Multi-Core CPUs. In: VLDB (2011)
6. Lemke, C., Sattler, K.-U., Faerber, F., Zeier, A.: Speeding up queries in column stores. In: Bach Pedersen, T., Mohania, M.K., Tjoa, A.M. (eds.) DAWAK 2010. LNCS, vol. 6263, pp. 117–129. Springer, Heidelberg (2010)

7. MacNicol, R., French, B.: Sybase IQ Multiplex - Designed For Analytics. In: VLDB (2004)
8. Mühe, H., Kemper, A., Neumann, T.: Executing Long-Running Transactions in Synchronization-Free Main Memory Database Systems. In: CIDR (2013)
9. Plattner, H.: A Common Database Approach for OLTP and OLAP Using an In-Memory Column Database. In: SIGMOD (2009)
10. Psaroudakis, I., Scheuer, T., May, N.: Task Scheduling for Highly Concurrent Analytical and Transactional Main-Memory Workloads. In: ADMS in Conjunction with VLDB (2013)
11. Schwalb, D., Faust, M., Krueger, J., Plattner, H.: Physical Column Organization in In-Memory Column Stores. In: Meng, W., Feng, L., Bressan, S., Winiwarter, W., Song, W. (eds.) DASFAA 2013, Part II. LNCS, vol. 7826, pp. 48–63. Springer, Heidelberg (2013)
12. Sikka, V., Färber, F., Lehner, W., Cha, S.K., Peh, T., Bornhövd, C.: Efficient Transaction Processing in SAP HANA Database - The End of a Column Store Myth. In: SIGMOD (2012)
13. Stonebraker, M., Abadi, D., Batkin, A., Chen, X., Cherniack, M., Ferreira, M., Lau, E., Lin, A., Madden, S., O'Neil, E.: C-store: A Column-oriented DBMS. In: VLDB (2005)
14. Willhalm, T., Popovici, N., Boshmaf, Y., Plattner, H., Zeier, A., Schaffner, J.: SIMD-Scan: Ultra Fast in-Memory Table Scan Using on-Chip Vector Processing Units. In: VLDB (2009)
15. Zukowski, M., Boncz, P., Nes, N., Heman, S.: MonetDB/X100—A DBMS in the CPU cache. IEEE Data Engineering Bulletin (2005)

A Functional Database Representation
of Large Sets of Objects

Ratko Orlandic[1], John Pfaltz[2], and Christopher Taylor[3]

[1] FairCom Corporation, Columbia, MO
[2] Dept. of Computer Science, University of Virginia
[3] Dept. Microbiology, Immunology and Parasitology, LSU Health Sciences Center,
New Orleans, LA

Abstract. This paper explores a novel way of implementing set-valued
operators that are used in analysis and retrieval in large social networks.
The software we describe has been implemented and thoroughly tested
in several demanding applications.

1 Introduction

Traditionally database systems have been optimized to retrieve those sets of objects which have specified attribute values. We might seek "all men in Queensland who play cricket". If an attribute is frequently used for retrieval, it may
be indexed. These traditional data sets are typically visualized as flat tables or
relations.

However, with the rise of social networks, we encounter a new kind of large
data set; one with "structure". It is not sufficient to just retrieve individual
object, or nodes, within the network; we may have to analyze both its local and
its global structure to find what we want [15]. Very often we want to analyze the
network's structural change over time [14].

One powerful approach to the understanding of network behavior is based on
operator theory, where an operator is a function mapping sets of elements into
other sets. Of particular interest have been closure operators. One such analytic
technique based on closure is described in Section 4. Writing efficient network
analysis software has required, first, a general method of set representation which
is briefly described in Section 2; and second, a highly effective way of implementing functional access. This, the major contribution of this paper, is described in
Section 3.

2 Set Representation

Perhaps the easiest, and most effective, way of representing any set Y is as
a bit string, in which an element $y \in Y$ if the corresponding bit is a 1. Let
$X = \{a, c, d, f\}$ and $Y = \{a, b, d, e\}$ with corresponding bit strings $X = [101101]$
and $Y = [110110]$. The logical *and*, or X&&Y, is $[100100] = \{a, d\} = X \cap Y$, and

H. Wang and M.A. Sharaf (Eds.): ADC 2014, LNCS 8506, pp. 189–197, 2014.

the logical *or*, X||Y, is [111111] = $\{a, b, c, d, e, f\}$ = $X \cup Y$. Subset comparison, $X \subseteq Y$ can then be expressed as "if X && Y = X".

These logical operators are $O(1)$. And even though longer bit strings may be needed to represent potentially unbounded element sets, in practice their theoretical linear performance becomes insignificant relative to other algorithmic operations. So effectively, these binary set operators are still $O(1)$ in practice.

Every universe is dynamic; as new objects are created, or encountered, they are entered into the universe and assigned a corresponding bit position, called its "index". If necessary, this integer index can be passed between procedures as a surrogate of the corresponding object. Set procedures, whether the binary operations of $\cup$, $\cap$, and $-$, or unary operations such as "insert x into Y" are all index based. Given an index integer we must be able to access the corresponding set element; and given a set element, we must be able to retrieve its index. The O-trees of the next section can do this rather nicely because they are key type independent.

All of the set-valued operator procedures described in Section 4 have been implemented as these kinds of templated C++ procedures operating over binary bit string representations of sets. It has been quite satisfactory, and we have tested our all codes with artificial sets of 20,000 elements.

3   O-Trees

A fundamental component of object-oriented databases is the ability to access an object, or set of objects, given a "key". The key may be unique, as an object name, or it may be an attribute shared by many objects. This access process is often called "information retrieval" and its associated literature is immense, *c.f.* [9]. We prefer the term "functional lookup" where the key is treated as the argument to the "lookup" function. Given an argument, or *key*, the function returns its corresponding *value*, which is often a pointer to an object. Basic techniques include linear arrays, when the argument is numeric, sequential search, hashing, tree search, among many others. Some are scalable; some are effective in distributed environments. In this section we describe a variant of B^+-tree search, called O-trees, which we have found to be very effective in the parallel execution of our functional database system, ADAMS [17].

O-trees, [11,13], behave much like the better known B^+-tree except that they have the following remarkable property. Any argument, which can be of arbitrary type and length (up to 254 bits, or 31 bytes), can be represented by a single 8 bit key.

An O-tree treats all operator arguments, or keys, as if they were just binary strings. Conceptually then, an O-tree is just a 0-complete trie[1] as shown in Figure 1, where its edges have been labelled with 0 or 1. A trie is said to be 0-*complete* if every 1-node has a corresponding 0-node. The empty dashed leaf of Figure 1

---

[1] The term "trie" (from "*retrie*val"), where a prefix search key is encoded by edge labels, was first coined by Fredkin [5]. Many variants have appeared in the literature.

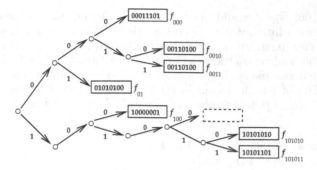

Fig. 1. A 0-complete trie

must be included in the trie to make it 0-complete. A node is a 1-node (0-node) if its incoming edge is labelled with a 1 (or a 0).

In Figure 1, seven functional values, f_{key} have been associated with 7 binary keys in such a way that these values constitute the leaves whose access paths are prefixes of their respective keys. A straightforward representation of such a binary trie would be most inefficient.

However, it can be shown that in any preorder traversal of a 0-complete tree, every leaf (except the last) must be followed by a 1-node. In the compact representation of an O-tree, the key of each leaf is denoted by the *depth* of the 1-node that follows it in the preorder sequence of the conceptual 0-complete tree, as shown in Figure 1. Observe, that while the leaf $(00110100, f_{0011})$ is at depth 4 in the trie of Figure 1, the entry in the index block is 2, the depth of the following 1-leaf $(01010100, f_{01})$. The last leaf which has no following 1-leaf is always associated with an imaginary depth 0. We let $D[\,]$ denote this sequence of depths in an index block.

Surprisingly, there is a very simple algorithm that uses this compact O-tree structure for search and retrieval. Given a key K, we begin any search by creating an integer vector $B[\,]$ that indicates the position of all 1 bits in K. For example

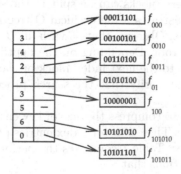

Fig. 2. The compact O-tree representation of the 0-complete trie of Figure 1

if $K = 00110100$, $B_K[\ ]$ would be $< 3, 4, 6, 9 >$. To ensure termination of the search algorithm, a final integer which exceeds the length of any key in the key space, in this case 9, must be appended to the sequence $B_K[\ ]$. In our actual code, we use 255, and restrict keys to be no longer than 254 bits.[2] The following simple algorithm compares the sequence $B_K[\ ]$ of one bits in the key against the sequence $D[\ ]$ of 1-node depths in the index block to determine which entry (returned as $depth\_j$) points to the leaf in which f_K will be found — if it exists at all.

```
procedure  SEARCH ( B, D, bit_i )
short      B[], D[], bit_i;
    // B is an array denoting the position of 1 bits in the key K,
    // D is the sequence of 1-node depths in the index block
    {
    int   depth_j;

    depth_j = 1;
    while (B[bit_i] <= D[depth_j])
        {
        if (B[bit_i] = D[depth_j])
            bit_i += 1;
        depth_j += 1;
        }
    return depth_j;
    }
```

The argument bit_i denotes which 1 bit in $B_K[\ ]$ is to be used to begin the comparisons in the block. Initially bit_i will be 1. Note that search through an index block using this procedure is sequential, as are many B^+-tree search procedures, except that the comparison at each entry is always between short 8 bit entries, regardless of the initial key type, or length.

The index block in the O-tree representation of Figure 2 has 8 entries. The 5th entry, with a *null* pointer, denotes the empty 0-leaf of Figure 1 which is needed to make it 0-complete. A more typical index block will have 200, or more, entries. But still, index blocks must have finite capacity. As the domain, or key space, of the function increases, these blocks can be split in the same manner a B^+-tree index blocks. Figure 3 illustrates a hierarchical O-tree representation over the same set of functional keys. Observe, that as with all B^+-trees, all leaves are at the same depth from the root. Depending on the depth of the O-tree, $depth\_j$ either denotes the pointer to the next index block or the desired leaf — usually an object. The reader may verify that the search procedure given above still works.

Using the 1 byte depth to compress the size of key entries in index blocks has an important consequence. These shorter index entries permit greater fan out at each level of the index tree. Table 1 compares the performance of B^+-trees and O-trees under the assumptions that

---

[2] This provides for 2^{254} distinct possible key, or argument, values.

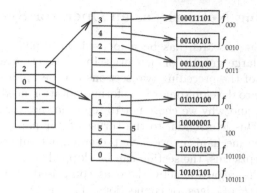

Fig. 3. A hierarchical O-tree representation of the 0-complete trie of Figure 1

Table 1. Comparison of B^+-trees and O-trees

| number | Number of items indexed | | | |
| of levels | B^+-tree | | O-tree | |
	Maximal	Expected	Maximal	Expected
1	256	177.4	409	334.8
2	65,536	45,414.4	167,281	94,915.8
3	16,777,216	8,056,514.5	68,417,929	26,908,629.3

(1) all index blocks are 2048 bytes in length,
(2) all pointers are 4 bytes, and
(3) B^+-tree keys are 4 byte integers, while O-tree depths are 1 byte shorts.

Using formulas found in [12], one can calculate The maximal, and expected, numbers of items indexed by a one level, two level and a three level B^+-tree and O-tree respectively is indicated. There is little reason to consider 4-level O-trees, because the expected 7,628,596,406.5 accessible nodes far exceeds the maximum 4-byte pointer value, which is $2^{32} = 4,294,967,296$. If the keys are multi-byte strings, the contrast between traditional B^+-trees and O-trees becomes more striking.

The biggest drawback of tree search for functional (or indexed) lookup versus other methods, such as extensible hashing [4,8], has always been the cost of descending through the tree. Even though its performance is theoretically $O(log\ n)$, pointer chasing is expensive, so that in practice linear $O(n)$ search is often preferable. That a three level O-tree with only small 2048 byte index blocks can be expected to handle 26.9 M functional $(arg, value)$ pairs makes this an effective technique for implementing functional lookup and operator evaluation. It has been thoroughly tested as the core of the object-oriented ADAMS database system [17], especially in parallel, distributed environments [10].

4   Representing Networks as an Operator System

Much of our current research has been focused on applying operator theory to the analysis of large graphs, or networks. In this section we show how the implemented code of the preceding two sections can be applied in practice. To do this we will have to develop some of the formal notation needed to explain our operator approach, and why we chose not to represent networks as large sparse matrices. None of this is essential to the thrust of our paper. Its sole purpose is to assert that the mechanisms we have presented actually have value. If the reader already accepts this, the section can be skipped.

It is customary to regard a network $\mathcal{N}$ as comprised of a set N of nodes, together with a set E of edges, or connections. That is, $\mathcal{N} = (N, E)$. Although this approach is quite viable, we have found it preferable to model networks in terms of the set N of nodes and a collection of operators. An **operator** is a single-valued function, say α, which for all subsets $Y \subseteq N$ yields a unique set $Y.\alpha \subseteq N$.[3] Thus operators map the subsets of N into N.

A fundamental operator is the domination operator, ρ, which is defined to be Y together with all nodes to which Y is "connected" or "related". The set $Y.\rho$ is said to be the **region dominated** by Y.[4] For all singleton sets $\{y\} \subset N$, $\{y\}.\rho$ is stored in the database. By definition,

$$Y.\rho = \bigcup_{y \in Y} \{y\}.\rho. \tag{1}$$

These sets, $Y.\rho$, can be either calculated using (1) or explicitly stored in the database as well and retrieved given the argument Y. Because the keys to O-tree retrieval are type independent, it is easy to provide either a set identifier Y, or the set itself, as the the argument to the ρ and η operators.

A subsidiary operator is the **neighborhood** operator, η, is defined by

$$Y.\eta = Y.\rho - Y. \tag{2}$$

Readily, this neighborhood operator is linked to a more traditional edge representation by $\{y\}.\eta = \{z | (y, z) \in E\}$ for all $y \in N$, or to a matrix representation by $\{y\}.\eta =$ the non-empty elements of row y. Observe that $Y.\eta \subset \cup_{y \in Y}\{y\}.\eta$, and so can not be easily calculated, as in (1). These sets, $Y.\eta$ and $Y.\rho$ have also been called the "open" and "closed" neighborhoods of Y, using a $N(Y)$ and $\bar{N}(Y)$ notation.

More importantly, given the network operators η and ρ, we can define a third operator, φ, by

$$Y.\varphi = Y \cup \{z \in Y.\eta | \{z\}.\eta \subseteq Y.\rho\}. \tag{3}$$

Readily, $Y.\eta \subseteq Y.\varphi \subseteq Y.\rho$. It is can be shown that φ is a **closure** operator [14]. Closure operators, which are: expansive ($Y \subseteq Y.\varphi$), monotone ($X \subseteq Y$ implies

---

[3]   For reasons unrelated to this paper, we prefer to denote set-valued operators using suffix notation rather than more usual prefix notation. Thus we write $\{y\}.\alpha$ instead of $\alpha(y)$.

[4]   This terminology is derived from the extensive "graph domination" literature [6].

$X.\varphi \subseteq Y.\varphi$), and idempotent ($Y.\varphi.\varphi = Y.\varphi$), are a central mathematical concept giving rise, for example, to notions of matroids (generalized independence) [1] antimatroids (concepts of convexity, acyclicity) [3], and greedoids (greedy algorithms) [7]. We use closure as a basic mechanism to study networks and their dynamic properties. In particular, "continuous" transformations of social networks can be defined in terms of closed sets [14].

Closure is not easily modelled using an edge set formalism. Yet, it can be fundamental to an understanding of the global connectivity properties of the network, and how they change. By way of example, the irreducible spine, $\mathcal{I}$, of a network $\mathcal{N}$, is one instance of the role closure can play in the analysis of large networks. One might expect every node of a network to be closed, that is $\{y\}.\varphi = \{y\}$. But, this is seldom the case. However, for all networks $\mathcal{N}$, there exists a unique sub-network $\mathcal{I}$ in which for all $y' \in \mathcal{I}$, $\{y'\}.\varphi' = \{y'\}$, where φ' is φ restricted to $\mathcal{I}$, or $\varphi|_{\mathcal{I}}$. We say $\mathcal{I}$ is **irreducible**. Moreover, there exists a procedure $\mathcal{R}$ such that $\mathcal{N} \xrightarrow{\mathcal{R}} \mathcal{I}$. $\mathcal{R}$ is a "continuous" transformation [14]. Pseudo-code for the key loop of this reduction process is:

```
for_each y in N {
    for_each z in y.nbhd {
        if (z.nbhd subset_of y.dom {
            remove (z); } } }
```

O-trees can provide rapid access to the sets $\{z\}.\eta$ and $\{y\}.\rho$; the *subset_of* operation is a bit string comparison. As shown in [16], the entire process is effectively $O(n)$, *i.e.* the first loop over all nodes, N, but in worst case $O(n^2)$.

Besides being unique (upto isomorphism) for every network $\mathcal{N}$, three important properties of $\mathcal{I}$ are that: (1) global connectivity is preserved by $\mathcal{R}$, that is a path $< x, \ldots, y, \ldots, z >$ from x to z through y exists in $\mathcal{N}$ if and only if a path $< x', \ldots, y', \ldots, z' >$, where $x' = x.\mathcal{R}$, $y' = \{y\}.\mathcal{R}$ and $z' = z.\mathcal{R}$, exists in $\mathcal{I}$ [15]; (2) if y is a "center" of $\mathcal{N}$ with respect to either "distance" or "betweenness" [2], then $y \in \mathcal{I}$; and (3) $\mathcal{I}$ is comprised of chordless cycles of length ≥ 4, which themselves have interesting properties [16]. Figure 4 illustrates such an irreducible core $\mathcal{I}$ (solid lines) of a 200 node network with 320 undirected connections and 6 connected components. The remaining reducible nodes are in a lighter font and are connected with dashed lines. Using (3), we observe that $\{161\}.\varphi = \{10, 22, 58, 59, 83, 93, 100, 108, 174\}$ in $\mathcal{N}$, but $\{161\}.\varphi = \{161\}$ in $\mathcal{I}$. Most of the networks we have been analyzing have several thousands of nodes, so as networks go, 200 nodes is rather small, but it is near the limit of what can be effectively illustrated.

So, a functional set-valued operator view of large networks can be quite rewarding. However, our purpose is not to justify this statement, but rather show how a set-valued functional database can be employed in practice.

5   Conclusions

A operator approach to graph and network analysis in terms of closure operators can yield interesting formal results as suggested in Section 4. But, to yield

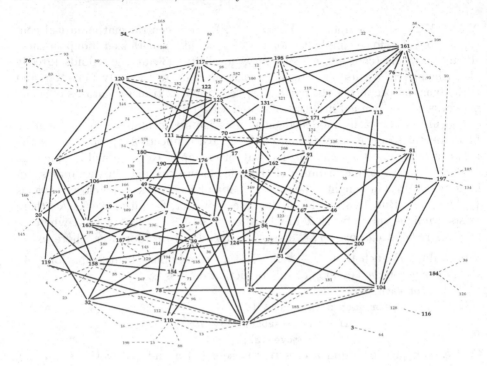

Fig. 4. The irreducible 53 node core of a 200 node network

computational results, it must be supported by highly effective method of functional lookup and set manipulation. In Sections 2 and 3 we have described two implementations that we have found to work rather efficiently.

However, we believe the functionality of O-tree indexing has not yet been fully explored. Although we have not worked with networks of more than 5,000 nodes, they are well within the capacity of a three-level O-tree according to Table 1. Nevertheless, we can easily imagine even larger networks that would require keys of length greater than 254 bits, and more than 30,000 expected objects. Expanding these capabilities appears to be little more than a coding exercise; yet it should be undertaken.

The objects of our research are all well-structured; but there is considerable interest in semi-structured information retrieval. The fact that O-tree retrieval is key type independent makes it a prime candidate for indexing such semi-structured data. However, we have had no experience with this. It remains a potential area of future research.

Acknowledgements. The authors would acknowledge the very helpful suggestions of three unknown referees.

References

1. Bonin, J.E., Oxley, J.G., Servatius, B. (eds.): Matroid Theory. Contemporary Mathematics, p. 197. Amer. Math. Soc., Providence (1995)
2. Dekker, A.: Conceptual Distance in Social Network Analysis. J. of Social Structure 6(3), 1–31 (2006)
3. Edelman, P.H., Jamison, R.E.: The Theory of Convex Geometries. Geometriae Dedicata 19(3), 247–270 (1985)
4. Flajolet, P.: On the performance evaluation of extendible hashing and trie searching. Acta Info. 20, 345–369 (1983)
5. Fredkin, E.: Trie Memory. Comm. ACM 3(6), 490–499 (1960)
6. Haynes, T.W., Hedetniemi, S.T., Slater, P.J.: Fundamentals of Domination in Graphs. Marcel Dekker, New York (1998)
7. Korte, B., Lovász, L., Schrader, R.: Greedoids. Springer, Berlin (1991)
8. Kumar, V.: Concurrent operations on extendible hashing and its performance. Comm. ACM 33(6), 681–694 (1990)
9. Manning, C.D., Raghavan, P., Schültz, H.: Introduction to Information Retrieval. Cambridge Univ. Press (2008)
10. Mikesell, D.R., Emanuel, W.R.: Interfacing an Object-Oriented Database System to a Global Primary Productivity Simulation. In: 12th Intern'l. Scientific and Statistical Database Conf., Berlin, Germany (July 2000)
11. Orlandic, R.: Design, Analysis and Applications of Compact 0-Complete Trees. Ph.D. Dissertation, Univ. of Virginia (May 1989)
12. Orlandic, R., Mahmoud, H.: Storage Overhead of O-trees, B-trees, and Prefix B-trees: A Comparative Analysis. Int. J. of Foundations of Computer Science 7(3), 209–226 (1996)
13. Orlandic, R., Pfaltz, J.L.: Compact 0-Complete Trees. In: Proc. 14th VLDB Conf., Long Beach, CA, pp. 372–381 (August 1988)
14. Pfaltz, J.L.: Mathematical Continuity in Dynamic Social Networks. In: Datta, A., Shulman, S., Zheng, B., Lin, S.-D., Sun, A., Lim, E.-P. (eds.) SocInfo 2011. LNCS, vol. 6984, pp. 36–50. Springer, Heidelberg (2011)
15. Pfaltz, J.L.: Finding the Mule in the Network. In: Alhajj, R., Werner, B. (eds.) Intern. Conf. on Advances in Social Network Analysis and Mining, ASONAM 2012, Istanbul, Turkey, pp. 667–672 (August 2012)
16. Pfaltz, J.L.: The Irreducible Spine(s) of Discrete Networks. In: Lin, X., Manolopoulos, Y., Srivastava, D., Huang, G. (eds.) WISE 2013, Part II. LNCS, vol. 8181, pp. 104–117. Springer, Heidelberg (2013)
17. Pfaltz, J.L., French, J.C.: Scientific Database Management with ADAMS. Data Engineering 16(1), 14–18 (1993)

Real-Time Exploration of Multimedia Collections

Juraj Moško, Tomáš Skopal, Tomáš Bartoš, and Jakub Lokoč

Charles University in Prague, Faculty of Mathematics and Physics, SIRET Research Group
Malostranské nám. 25, 118 00 Prague, Czech Republic
{mosko,skopal,bartos,lokoc}@ksi.mff.cuni.cz
http://www.siret.cz

Abstract. With the huge expansion of smart devices and mobile applications, the ordinary users are consistently changing the conventional similarity search model. The users want to explore the multimedia data, so the typical query-by-example principle and the well-known keyword searching have become just a part of more complex retrieval processes. The emerging multimedia exploration systems with robust back-end retrieval system based on state of the art similarity search techniques provide a good solution. They enable interactive exploration process and implement exploration queries tightly connected with the user interface. However, they do not consider larger response times that might occur. To overcome this, we propose a scalable exploration system **RTExp** that allows evaluating the similarity queries in the near real time depending on user preferences (speed / precision). We describe building parts of the system and discuss various real-time characteristics for the exploration process. Also we provide results from the experimental evaluation of time-limited similarity queries and corresponding exploration operations.

1 Introduction

The continuously growing multimedia collections provide generally unstructured data, so it is almost impossible to completely understand what kind of information these datasets contain and which data might be relevant for individual users. Since it is difficult to index such data or to query it using the traditional approaches (e.g., full-text indexes), new techniques employing the *similarity search* paradigm [21] have emerged.

The similarity search concept applies the content-based comparison of multimedia objects driven by object similarities. This approach typically leads to *query-by-example* searching in which a user specifies the query with an initial object with additional conditions. However, this type of *targeted search* (or controlled querying), with a sample query object is not intuitive enough when compared to real user expectations.

In specific scenarios, the user does not really know what (s)he is searching for, so (s)he cannot provide the appropriate query object. Instead, the user iteratively *browses* or *explores* the database until spotting the right object(s). Here, we talk about *indirect searching*. Recently, there appear many approaches to the interactive indirect search for multimedia collections with the main focus put on *multimedia exploration systems* [14,17], that aims at more sophisticated and faster kind of browsing.

We consider the *exploration* as a user-controlled (interactive) process of viewing and browsing the multimedia collections in a way which is arbitrary, unpredictable, and not

H. Wang and M.A. Sharaf (Eds.): ADC 2014, LNCS 8506, pp. 198–205, 2014.
© Springer International Publishing Switzerland 2014

predefined at all. Initially, the system displays and visualizes a very small portion of the underlying dataset. The user can pick an object, click on or zoom in/out specific parts of the currently visualized items, so that the search becomes partly targeted during the interactive search process.

The recently proposed engines [14,17,2] encounter real problems with processing very large databases (millions of objects or more), because they do not consider the most important challenge – to guarantee the worst query performance for arbitrarily large collections which is the main objective of this paper. Unlike targeted querying, in which the user accepts short delays for the query execution, during exploration the process has to be smooth and instant. Hence, for multimedia exploration the underlying routines have to operate in near real time to provide the illusion of a smooth process.

2   Related Work and Motivation

The concept of multimedia exploration systems has been recently introduced as the combination of two major factors: (1) browsing multimedia collections, and (2) using approximate similarity search [2,14,17]. Each of the systems introduces some novel ideas such as the intuitive interface to a visualized collection [17], or the separated middle layer in the system architecture [16].

The approximate similarity search arises from the idea of exchanging the query effectiveness (precision) for the query efficiency (speed). Users adjust this trade-off at/before the query time using various approximation parameters, for example a probability of an error in the query result [22,6], the maximum number of similarity/distance computations allowed, or the threshold on the improvement of the query result [22].

The classification of approximate search scenarios [15] reveals different principles and organizes them in a consistent way. One of the inspiring approaches is the RC_{ES} which stands for *reducing comparisons* and *early termination*. Another relevant technique is the concept of incremental similarity search [20]. It works with tree index structures and assigns higher priority to more promising nodes. Based on this priority, we traverse unprocessed nodes within the tree and stop the process whenever the most promising node in the processing queue guarantees worse values than currently the "worst" node in the result set (e.g., smaller similarity than the kth nearest neighbor). This technique was later proven to be range-optimal [11] and also optimal for disk page accesses [4]. One of the first adoptions of this idea has been in spatial databases [10] and afterwards also in metric indexes [11].

Although we could utilize any of these systems, they have to be adopted for different purpose (multimedia exploration) than originally designed (similarity queries). As the crucial requirement on the multimedia exploration process is a smooth navigation in the collection, the indexes behind providing partial similarity queries have to guarantee real-time responses, either given a real-time constraint for execution, or even anytime query termination when no time constraint is available. If such a guarantee is satisfied, the entire exploration system could be designed to guarantee smooth navigation of the user in the collection, where the underlying streams of similarity queries are scheduled and executed such that the total real-time constraint is kept.

2.1 Multimedia Exploration in Mobile Devices

There is no doubt that classical computers and notebooks are being dominated by small, smart, touch-based mobile devices constantly connected to the Internet. As a consequence, users can search, browse, download/upload multimedia data practically anytime and anywhere. On the other side, small displays and limited control options represent new challenges for mobile multimedia applications trying to provide the multimedia data in the most convenient and also entertaining way such as multimedia exploration can offer [2,18].

Although multimedia exploration techniques apply index support and approximate search to improve the efficiency of whole process, to the best of our knowledge, none of the related methods considers the real-time responses of the exploration process in the presence of huge multimedia datasets and mobile environment. We believe that guaranteeing a real-time response is one of the most crucial properties of a successful multimedia exploration system, even more important than the steadily high precision of the retrieval. Hence, we focus on approximate search techniques guaranteeing user-defined response time and allowing occasional inaccuracies for long running queries. Furthermore, the response time can be controlled by users who decide the trade-off between the precision and the efficiency. For example, for a casual exploration of an unknown collection the user prefers fast response times, but when searching for some specific objects, the user will wait longer for a more qualitative result.

2.2 Real-Time Similarity Queries

Before we introduce our system, we provide necessary background for similarity queries performed in limited time frames to get results near real-time. The most popular queries are the k nearest neighbors (kNN) queries that return k most similar objects to the query object [21]. Our main intention for similarity exploration queries is to limit the response time. So, we define kNN(t) as the timely limited kNN query which returns up to k most similar objects obtained from database objects accessed by the similarity index during query processing within the restricted time frame t. These queries occur in batches denoted as the *query stream* initiated as the result of user interaction with the system.

3 Implementing RTExp System

In this section, we provide the detailed information about the proposed **RTExp** (Real-Time Exploration) system and describe the data flow with typical use cases. The architecture consists of *presentation*, *logic*, and *data* layer with one extra layer which guarantees inter-layer communication and proper data flow between individual components (see Fig. 1a). The following list outline characteristics of individual layers:

1. The presentation layer represents intuitive and comfortable graphical user interface. Each user action to a visualized exploration space is transformed into the sequence of *exploration operations* and the corresponding query stream at the logic layer.

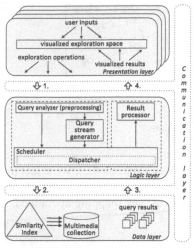

(a) High-level architecture   (b) iPAD prototype

Fig. 1. RTExp System overview

2. The logic layer performs two main tasks: (1) translate user exploration operations into query streams, and (2) deliver partial/final results from the data layer back to the user. First, we decide whether (a) we terminate the currently running query stream, or (b) we create a new query stream (*Query stream generator*). The *Scheduler* schedules and controls all requests with a prioritized queue based on user settings, while the *Dispatcher* transfers query streams to be executed and the *Result processor* decides which query results are propagated to the presentation layer.

3. Data layer consists of the *multimedia collection* with a dedicated *similarity index*. We evaluate and compare several metric access methods (MAMs) [21], results are provided in Section 5. In the future, we also plan to involve non-metric access methods [19,1]. Two factors influence real-time capabilities and system performance:

   (a) *different query execution plans* for computationally *cheap* (executed "as is") and *expensive* (approximated in some way) similarity queries. For example, in M-tree [7] we use the top-down search strategy for cheap queries, however for the expensive ones we apply the bottom-up approach (fast traversing to the leaf level followed by enhancing the first results).

   (b) *instant evaluation* of similarity queries (*instant queries*). The benefits of instant queries can be utilized when (a) preferring quick results over the relevancy, (b) terminating obsolete queries, or (c) when a rapid increase in the number of simultaneous queries occurs.

4   Exploration Operations and User Interface

The more intuitive and ergonomic interface the client application provides, the more likely the end users will use the system and benefit from it. Inspired by the similarity-based layout approaches [14], we implement two most important user actions in the

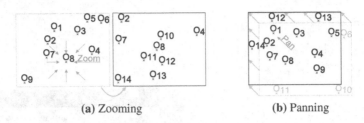

(a) Zooming   (b) Panning

Fig. 2. Exploration user actions

exploration process – *zooming* (Section 4.1) and *panning* (Section 4.2). We select these two actions as ideal candidates for basic exploration operations because they are intuitive and have been known to billions of Internet users as web mapping operations.

4.1 Zooming

Imagine we currently have a visualized set of items from the multimedia collection represented by objects O_i (see Fig. 2a). In this case, we are interested in the context of objects similar to the specific object O_8, so we *zoom in* to this object. The arrows show the zooming target which subsequently reveals previously invisible objects O_{10}, O_{11} ...O_{14}. The newly discovered objects are typically more similar than the previously shown objects or provide the visualization of more specific object clusters. On the contrary, if we apply the opposite action of *zooming out*, we get new objects that are less similar to the previous ones or we obtain more general object clusters.

Whenever a user executes a zooming action, we create the adequate exploration operation that consists of (a) the target zooming point, (b) the object closest to the target zooming point, and (c) objects closest to the boundaries. *Zoom in* initiates a single similarity query for which the query object is the closest object to the zooming target point, while for *zoom out* the query stream consists of multiple similarity queries for which the query objects are closest objects to the bounding rectangle.

4.2 Panning

This action advances the exploration in the specific direction (see Fig. 2b). The dotted rectangle represents the state before we apply *panning*, the solid one displays the new state, while the arrows outline the panning direction.

For panning actions, we create the appropriate exploration operation which includes several parameters such as (a) panning direction and its volume, (b) objects closest to visualized exploration space boundaries following the panning direction, and (c) objects that get outside the new visualized exploration space boundaries. Then, we initiate the query stream of similarity queries for which the query objects are objects closest to boundaries of the visualized exploration space following the panning direction.

4.3 RTExp Presentation Layer for Mobile Devices

To validate the feasibility of the proposed exploration system, we implemented the first prototype (called RTExp) with the user interface for iPad tablets (see Fig. 1b). The

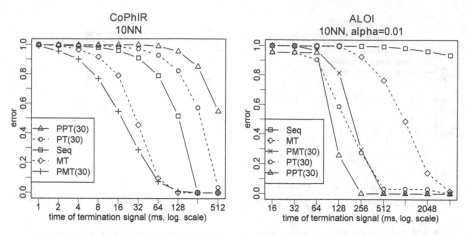

Fig. 3. *k*NN queries on the CoPhiR and ALOI database – inter-MAM comparison

powerful slider control at the bottom enables the user to adjust the exploration process. Sliding to the left gives more precise (but potentially slower) results, while putting the slider to the right returns faster (yet not necessary relevant) results.

5   Experimental Evaluations

Initially, we study the performance of RTExp system and depict the comparison of all evaluated MAMs: M-tree (*MT*), PM-tree (*PMT*), pivot tables (*PT*) and ptolemaic pivot tables (*PPT*) on two multimedia collections (see Fig. 3). While for subset of CoPhiR dataset [5], consisting of 1,000,000 images with cheap distance function, PM-tree is a clear winner (and pivot tables, behave even worse than the sequential scan (*Seq*)), the ALOI database [9], composed of 70,000 images with expensive SQFD [3], suits better for pivot tables. We also see that simple implementations of instant query termination may not be sufficient – the fastest index on ALOI reaches some reasonable results after 100ms. Hence, we need to design more specific approaches to instant query termination in the future.

Based on the previous experimental results, we employ instant query termination into simulation of real user *exploration scenario* (see Fig. 4). As the data layer, we use PM-tree, the best approach from the previous experiment. From results we can observe the major benefit of instant query termination - the constantly low execution time per a single exploration operation. Hence, the exploration process is really smooth, while the execution of the complete non-terminated operations would lead to unwanted delays.

Next, we study the precision error of exploration operations, each consists of several similarity queries. We join results from all (terminated) queries and compute the precision error with Jaccard distance [21] by comparing these results with joined results of complete (non-terminated) queries. The conclusion of this experiment is that the error correlates with the time of non-terminated query in case when all queries have the same amount of time for evaluation. Therefore, a *query analyzer* and/or modification of query evaluation is really necessary to improve the precision.

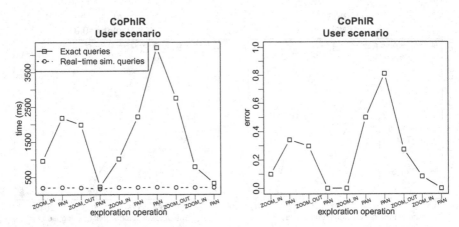

Fig. 4. User exploration scenario - time and precision error of exploration operations

6　Challenges and Future Work

During the development of RTExp system prototype, we experienced several issues that we would like to highlight, suggest possible solutions, and address them afterwards.

- **Continuous query evaluation.** For continuous performance, we intend to adapt the *publish/subscribe data delivery* [13]. Whenever a user executes an action, the visualized exploration space is continuously updated with partial/final results, while the system is evaluating the query stream in the background (*Result processor*).

- **Visualization of the multimedia collection.** We do not address the nontrivial problem of mapping the multimedia collection to the visualization space. We use the physical model based on the multidimensional scaling and the simulated annealing [8], successfully used and verified in another exploration system [12]. Later, we intend to employ more sophisticated solution such as the one suggested in [14].

Besides the already mentioned issues, in the future we want to focus on improving the precision of returned results while keeping the evaluation time at the same or even shorter values (e.g., employing data annotations).

7　Conclusion

Driven by requirements of end users, such as the modern user interface or truly continuous exploration, we propose the real-time multimedia exploration system RTExp that meets these criteria by providing scalable multi-layered architecture, applying real-time similarity exploration queries, and delivering intuitive user interface. Besides the description of the proposed system, we also verify our ideas in practice by developing a functional prototype with a user interface targeted for mobile devices and evaluating the system performance which demonstrates the viability of our multimedia exploration system.

Acknowledgments. This research has been supported in part by Czech Science Foundation project 202/11/0968 and by Grant Agency of Charles University projects 567312 and 910913.

References

1. Bartoš, T., Skopal, T., Moško, J.: Towards Efficient Indexing of Arbitrary Similarity. SIGMOD Record 42(2), 5–10 (2013)
2. Beecks, C., Skopal, T., Schöffmann, K., Seidl, T.: Towards large-scale multimedia exploration. In: DBRank 2011, Seattle, WA, USA, pp. 31–33 (2011)
3. Beecks, C., Uysal, M.S., Seidl, T.: Signature quadratic form distance. In: Conference on Image and Video Retrieval, CIVR 2010, pp. 438–445. ACM, New York (2010)
4. Böhm, C., Berchtold, S., Keim, D.A.: Searching in high-dimensional spaces: Index structures for improving the performance of multimedia databases. ACM 33(3), 322–373 (2001)
5. Bolettieri, P., Esuli, A., Falchi, F., Lucchese, C., Perego, R., Piccioli, T., Rabitti, F.: CoPhIR: A test collection for content-based image retrieval. CoRR abs/0905.4627v2 (2009)
6. Chávez, E., Navarro, G.: A probabilistic spell for the curse of dimensionality. In: Buchsbaum, A.L., Snoeyink, J. (eds.) ALENEX 2001. LNCS, vol. 2153, pp. 147–160. Springer, Heidelberg (2001)
7. Ciaccia, P., Patella, M., Zezula, P.: M-tree: An Efficient Access Method for Similarity Search in Metric Spaces. In: VLDB 1997, pp. 426–435. Morgan Kaufmann Publishers Inc. (1997)
8. Fruchterman, T.M.J., Reingold, E.M.: Graph drawing by force-directed placement. Softw. Pract. Exper. 21(11), 1129–1164 (1991)
9. Geusebroek, J.M., Burghouts, G.J., Smeulders, A.W.M.: The amsterdam library of object images. International Journal of Computer Vision 61(1), 103–112 (2005)
10. Hjaltason, G.R., Samet, H.: Distance browsing in spatial databases. ACM Trans. Database Syst. 24(2), 265–318 (1999)
11. Hjaltason, G., Samet, H.: Incremental Similarity Search in Multimedia Databases. Computer science technical report series, University of Maryland (2000)
12. Lokoč, J., Grošup, T., Skopal, T.: Image exploration using online feature extraction and reranking. In: ICMR 2012, pp. 66:1–66:2. ACM, New York (2012)
13. McIlvride, B.: Nine core requirements for real-time cloud systems (January 2012), http://real-timecloud.com/2012/01/12/nine-core-requirements-for-real-time-cloud-systems
14. Nguyen, G.P., Worring, M.: Interactive access to large image collections using similarity-based visualization. Journal of Visual Languages and Computing 19(2), 203–224 (2008)
15. Patella, M., Ciaccia, P.: Approximate similarity search: A multi-faceted problem. J. of Discrete Algorithms 7(1), 36–48 (2009)
16. Santini, S., Jain, R.: Integrated browsing and querying for image databases. IEEE Multimedia 7(3), 26–39 (2000)
17. Schaefer, G.: A next generation browsing environment for large image repositories. Multimedia Tools and Applications 47, 105–120 (2010), doi:10.1007/s11042-009-0409-2
18. Schoeffmann, K., Ahlstrom, D., Beecks, C.: 3D Image Browsing on Mobile Devices. In: ISM 2011, pp. 335–336. IEEE Computer Society, Washington, DC (2011)
19. Skopal, T.: Unified framework for fast exact and approximate search in dissimilarity spaces. ACM Trans. Database Syst. 32(4) (November 2007)
20. Uhlmann, J.K.: Implementing Metric Trees to Satisfy General Proximity/Similarity Queries (1991) manuscript
21. Zezula, P., Amato, G., Dohnal, V., Batko, M.: Similarity Search: The Metric Space Approach. Springer (2005)
22. Zezula, P., Savino, P., Amato, G., Rabitti, F.: Approximate similarity retrieval with m-trees. The VLDB Journal 7(4), 275–293 (1998)

XEdge: An Efficient Method for Returning Meaningful Clustered Results for XML Keyword Search[*]

Wenxin Liang[1],[**], Yuanyuan Gan[2], and Xianchao Zhang[3]

School of Software, Dalian University of Technology, Dalian 116620, China
{wxliang,xczhang}@dlut.edu.cn, roundcircle@foxmail.com

Abstract. In this paper, we investigate the problem of returning meaningful clustered results for XML keyword search. We begin by presenting a multi-granularity computing methodology, in order to make full use of the structural information of XML trees to extract features. In this method, we first propose the concept of Cluster Compactness Granularity (CCG) to partition the search results into different clusters, which enable users to precisely and quickly seek their desired answers, according to the connection compactness between LCA nodes. We then propose the concept of Subtree Compactness Granularity (SCG) to rank individual results within clusters and measure the query result relevance. Furthermore, we define a novel semantics of Compact LCA (CLCA), which not only improves the accuracy by eliminating redundant LCAs that do not contribute to meaningful answers, but also overcomes the shielding effects of SLCA-based methods. Using the proposed CCG and SCG features and the CLCA semantics, we finally implement an efficient algorithm called XEdge for generating meaningful clustered results. Comparing with the existing methods such as XSeek and XK-LUSTER, the experimental results demonstrate the effectiveness of the proposed multi-granularity clustering methodology and validity of the complemented ranking strategy, as well as the meaningfulness of CLCA semantics.

1 Introduction

Keyword search is a proven user-friendly way to query XML databases, since users do not have to learn query language that convey semantic meanings, and can process queries without any prior knowledge about the structure of underlying data. Numerous studies [1] [2] [3] [4] [5] and [6] on XML keyword search have been done to retrieve results from large document collections. However, most of them are focusing on proximate keyword search and the results obtained are far from satisfactory. Therefore, performing keyword searches over XML documents results in such issues as follows.

Meaningless Results. Existing methods usually return answers that are either irrelevant to the user's intention, or not meaningful or informative enough. In addition, since XML keyword search engines return XML snippets as final answers, there are

---

[*] This work was partially supported by NSFC (No. 61272374, 61300190), Program for NCET in University of China (No. NCET-11-0056), Specialized RFDP of Higher Education (No.20120041110046), Key Project of Chinese Ministry of Education(No. 313011) and the Fundamental Research Funds for the Central Universities (No. DUT13JR04).
[**] Corresponding author.

H. Wang and M.A. Sharaf (Eds.): ADC 2014, LNCS 8506, pp. 206–213, 2014.

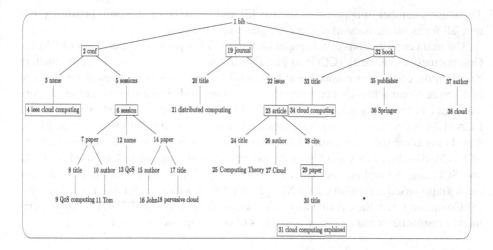

Fig. 1. Example XML tree

definitely overwhelming results containing many meaningless ones. Figure 1 shows an example XML tree, in which the numbers are node IDs sorted by pre-order sequence, and the nodes circled by rectangles are the candidate LCAs(Lowest Common Ancestors). Considering the keyword query $Q = "cloud\ computing"$, $session(6)$ should be a meaningless result of this query, because matching nodes (9) and (18) do not belong to the same paper.

Shielding Effect. Existing SLCA(Smallest Lowest Common Ancestor)-based approaches [2] proposed various strategies for identifying relevant matches by disqualifying certain SLCA nodes. The basic idea is that a candidate LCA node is considered to be optimal if it is relatively lower. Considering the query Q again, node (23) and (31) which contain the both keywords should be the correct answers. However, the result should be the lower subtree rooted at node (31), which omits upper one rooted at $article(23)$. Therefore, the SLCA-based methods will yield shielding effects on upper LCA nodes, which discard some relevant answers. Accordingly, how to obtain the meaningful results and overcome the shielding effects of SLCA-based methods becomes urgent.

Results Clustering. Consider the query Q with multiple searching semantics, such as: (i) find conferences about cloud computing; (ii) find papers whose titles contain cloud computing; (iii) find books about cloud computing; $(iiii)$ find papers about computing written by Cloud. Users may be interested in one or more of these interpretations. Since simple list of results is not helpful in this case, returning clustered results can be adopted to address this issue from a different perspective.

Result Ranking. There are multiple candidate answers in one cluster. How to measure the relevance of each search intention within clusters and rank these individual results based on relevance are challenging. The subtree rooted at $article(23)$ should be returned as a result because the search intentions of users are diverse. However, such kind of results should be given a lower rank, as they are not the major search intentions.

There are a number of approaches such as [3] and [6] to compute result proximity, but they all focus on the issue of ranking the generated clusters.

The main contributions of this paper include: (1) We propose the concept of Cluster Compactness Granularity (CCG) to partition the search results into different clusters based on the connection compactness between LCA nodes. (2) We present the concept of Subtree Compactness Granularity (SCG) to rank individual results within clusters and measure the query result relevance. (3) We define a novel semantics of Compact LCA (CLCA), which not only improves the accuracy by eliminating redundant LCAs that do not contribute to meaningful answers, but also overcomes the shielding effects of SLCA-based methods. (4) Using the proposed clustering algorithm based on CCG and SCG and the ranking mechanism based on CLCA, we finally implement an efficient graph-based algorithm called XEdge for generating meaningful clustered results. (5) Comparing with the existing methods such as XSeek and XKLUSTER, the experimental results demonstrate the effectiveness of our proposed methods.

2   Multi-granularity Methodology

2.1   Cluster Compactness Granularity (CCG)

Definition 1. *[Link Branch(LB)]A link branch is a top-down path from the top node to the bottom one. The length of Link Branch (LLB) between any two nodes equals to the amount of nodes in the path.*

In this paper, we utilize a simplified XPath-like representation to model LB. For example, the LB between $journal(19)$ and $article(23)$ in Fig. 1 is $journal(19)/issue(22)$ $/article(23)$, and obviously $LLB((19),(23)) = 3$.

Definition 2. *[Cluster Compactness Granularity(CCG)] Given two matching subtrees rooted at u and v ($u \neq v$), the cluster compactness granularity between them, $CCG_{u,v}$ is calculated by Equation 1, where $DEPTH(r)$ denotes the depth of a subtree's root node r (in particular, the depth of the root in a document tree is 1).*

$$CCG_{u,v} = \frac{LLB(u, LCA(u,v)) + LLB(v, LCA(u,v)) - 1}{DEPTH(LCA(u,v))} \qquad (1)$$

By calculating the CCG between any two LCA nodes in Fig. 1, we can obtain a CCG matrix as shown in Table 1. The CCG satisfies the actual connection relationship between any two LCA nodes. Therefore, we can cluster the results such that the CCG of results is as small as possible within the same cluster, and as large as possible in different clusters. In this way, we can finally obtain three clusters, $C_1 = \{(2),(4),(6)\}$, $C_2 = \{(19),(23),(31)\}$ and $C_3 = \{(32),(34)\}$, which represents three different matching semantics, "conference", "journal" and "book".

2.2   Subtree Compactness Granularity (SCG)

Definition 3. *[Size of Matching Subtree(SMS)] Given a matching subtree $T(r)$, the size of matching subtree $SMS(T(r))$ is defined as the total length of Link Branch from root r to each matching node m_i, which can be calculated by the following equation.*

Table 1. CCG Matrix(Preorder Sequence)

CCG \ Node IDs Node IDs	(2)	(4)	(6)	(19)	(23)	(31)	(32)	(34)
(2)	0	1.5	1.5	3	5	9	3	5
(4)	1.5	0	2.5	5	7	11	5	7
(6)	1.5	2.5	0	5	7	11	5	7
(19)	3	5	5	0	1.5	3.5	3	5
(23)	5	7	7	1.5	0	1.25	5	7
(31)	9	11	11	3.5	1.25	0	9	11
(32)	3	5	5	3	5	9	0	1.5
(34)	5	7	7	5	7	11	1.5	0

$$SMS(T(r)) = \sum_{i=1}^{n} |LLB(r, m_i)| \qquad (2)$$

Definition 4. *[Subtree Compactness Granularity(SCG)] For a matching subtree $T(r)$, the subtree compactness granularity (SCG) is defined by Equation 3 as the percentage of the size of $T(r)$, $SMS(T(r))$ out of the depth of the root of $T(r)$, DEPTH(r).*

$$SCG(T(r)) = \frac{SMS(T(r))}{DEPTH(r)} \qquad (3)$$

Equation 3 indicates that a matching subtree with smaller SCG means the result is more meaningful and should be given a higher rank. Given one LCA node l, it is obvious that more than one matching subtree rooted at l can be retrieved during traversing the XML tree. In order to clarify such matching subtrees, we present $SCGSet$ that denotes the various matching subtrees rooted at one LCA node. All the SCGSets calculated from Fig. 1 are shown in Table 2.

Table 2. Example SCGSets

LCA nodes	SCGSet	LCA nodes	SCGSet
conf(2)	{4, 4}	article(23)	{1.25, 1.75, 1.75}
ieee cloud computing(4)	{0.25}	cloud computing explained(31)	{0.125}
session(6)	{1.75}	book(32)	{2.5}
journal(19)	{3.5, 4.5}	cloud computing (34)	{0.25}

2.3   Combination of CCG and SCG

Given a set of LCA nodes, it is reasonable to cluster them based on CCG theory, such that the CCG value of any two LCA nodes in the same cluster is as small as possible. Within one cluster, the individual LCA node may have different SCGs which indicate different matching semantics. Take cluster $C_1 = \{(2), (4), (6)\}$ as an example, the subtree rooted at node (4) is top-ranked because of minimum SCG score with the semantics of "find the conference on cloud computing", which is considered to be a major search intention. As for node (2), it contains two matching subtrees representing minor search intentions, and hence they are given the lowest rank because of maximum SCG scores. However, node (6) is not a meaningful result and should be discarded from the result. To address this issue, we present a novel CLCA semantics in the next section.

2.4 Identifying CLCAs

Definition 5. *[Summary Tree(ST)] It can be modeled as a tree, and each node g is recognized as a group consist of a set of nodes denoted as $g.pid[]$. (1) $g.id$ is denoted by the node identifier of the first node in group g. (2) Each index value $k(k \geq 0)$ of $g.pid[]$ corresponds to one element node in group g, denoted as $g : k$. (3) $g.pid[k]$ points to one element node in group g_p which is the parent node of group g, that is, $g_p : g.pid[k]$ is the parent node of $g : k$.*

Definition 6. *[Compact LCA(CLCA)] Given a matching subtree $T(r)$ rooted at r and m matching nodes $n_1, n_2, \cdots, n_m$, $CLCA(n_1, n_2, \cdots, n_m) = r$, iff, for $\forall 1 < i \leq j \leq m, n_i.pid[0] = n_j.pid[0]$, that is, m matching nodes are homogenous.*

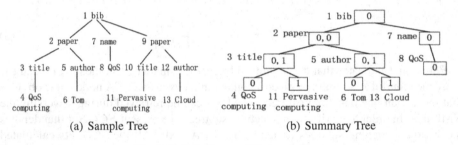

Fig. 2. Sample Tree and Summary Tree

An example ST shown in Fig. 2(b) is summarized from Fig. 2(a). For the query $Q = "cloud\ computing"$, $paper(9)$ is actually a CLCA node as its matching nodes satisfy that $(11).pid[0] = (13).pid[0] = 1$, and should be deemed meaningful results. However, $paper(2)$ and $paper(9)$ belong to different top-down paths, as a result, $(4).pid[0] \neq (13).pid[0]$. Thus, the LCA node $bib(1)$ is not considered to be a CLCA node and should be discarded.

3 XEdge Algorithm

3.1 Inferring GRCs

Definition 7. *[Graph-based Result Cluster (GRC)] Given an XML document tree T, suppose that V_{clca} denotes the set of CLCA nodes. The Graph-based Result Cluster GRC is defined as a connected directed graph $G(V, E)$ such that: 1) $V \subseteq V_{clca}$; 2) For any $(u, v) \in E, \forall v' \in V_{clca}(v' \neq v)$ such that $CCG_{u,v} \leqslant CCG_{u,v'}$.*

The basic idea is to infer GRCs while traversing data tree, and it operates like extracting shortest paths. The edges form many isolated GRCs during the generating process, and thus an individual connected GRC represents a unique cluster.

The calculation of GRCs starts from any CLCA node $u \in V_{clca}$, in each step we add the edge (u, v) with the minimum CCG score to the GRC until all the nodes $u \in V_{clca}$ have been traversed. Example GRCs extracted from Fig. 1 are shown in Fig. 3.

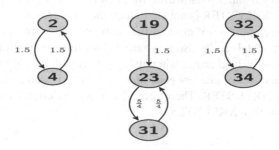

Fig. 3. Example GRCs extracted from Fig. 1

3.2 Algorithm XEdge

Using the proposed CCG and SCG features and the CLCA semantics, an efficient algorithm called XEdge for generating meaningful clustered results is implemented by the following steps. Given a keyword query Q on XML document tree T, we begin by identifying candidate matching subtrees. After that, massive matching subtrees $MT(l_i)$ ($l_i \in V_{lca}$) contained all LCA nodes will be returned. Subsequently, for each matching subtree $MT(l_i)$, we check whether l_i is a CLCA node based on Definition 6. If the node l_i is not a CLCA, this subtree will be removed from the result set, otherwise it will be used to calculate the SCG. Next, we calculate CCG between any two CLCAs and extract GRCs based on Definition 7. Then, the extracted GRCs can be ranked by using the SCG features, and we finally obtain the ranked result clusters, which not only improves the search accuracy, but also overcomes the shielding effects of SLCA-based approaches.

4 Experimental Evaluation

We compared the XEdge algorithm with XSeek [2] and XKLUSTER [3] using three real datasets: DBLP, Mondial and SigmodRecord downloaded from [7]. All algorithms were implemented in Java and run on a 2.93GHz Pentium(R) Dual-Core machine with 2GB RAM running Windows 7. We selected 7 keyword queries with less than 5 terms for each dataset. We employed three metrics, the number of clusters, precision and recall to evaluate all the algorithms.

4.1 Number of Clusters

We evaluated the number of clusters generated by XEdge and XKLUSTER using Sig-modRecord dataset. Since XKLUSTER is a distance threshold-dependent algorithm, we used different distance threshold (0, 1.0, 2.0, 4.0, 9.0) in the experiments. The experimental results are shown in Fig. 4. For such queries containing less keywords as $S3$ and $S4$, XKLUSTER generates almost the same amount of results as XEdge does when the threshold equals to 4.0 and 9.0. However, for such queries containing more keywords as $S1$, $S6$, and $S7$, XKLUSTER is not enough efficient compared with XEdge because it generates too much number of clusters that may include more meaningless results. Besides, the experimental results also illustrate that the number of clusters generated by XKLUSTER is sensitively changed when using different threshold even for the same query. While, the number of clusters generated by XEdge is stable and much less than those generated by XKLUSTER. Therefore, XEdge is more effective and adaptable for the real applications than XKLUSTER.

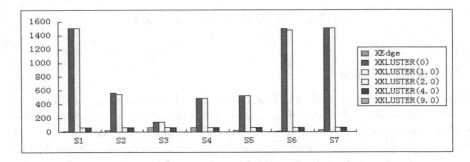

Fig. 4. The comparison between XKLUSTER and XEdge

4.2 Effectiveness of Methodology

To evaluate the effectiveness of the proposed method, we compared XEdge with XK-LUSTER and XSeek using precision and recall. The experimental results demonstrate that XEdge outperforms the two baseline methods significantly in terms of precision and recall. From Fig. 5(a) and 6(a), XEdge achieves both high recalls and precisions on Mondial and DBLP datasets in most queries. XKLUSTER achieves high precision except for the queries $M4 \sim M7$, $D3$, $D4$ and $D6$ containing more than two keywords, because XKLUSTER utilizes the "OR" logic in these queries, which causes more false-positives. XSeek achieves low precision in queries $M1$, $M2$, $D1$, $D2$, $D6$ and $D7$, because it has shielding effect towards higher LCA nodes. As observed from Fig. 5(b) and 6(b), XKLUSTER achieves high recalls in almost all the given queries. This is because XKLUSTER supports "OR" logic, namely, it can capture more semantics and hence possibly outputs more meaningful results. However, as analyzed before, the high recalls may result in low precisions especially when the queries containing more keywords.

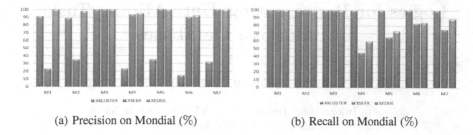

(a) Precision on Mondial (%)   (b) Recall on Mondial (%)

Fig. 5. Measures on Mondial dataset for the keyword queries

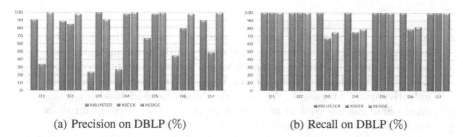

(a) Precision on DBLP (%)   (b) Recall on DBLP (%)

Fig. 6. Measures on DBLP dataset for the keyword queries

5   Conclusion

We have presented a multi-granularity methodology called XEdge, which makes full use of the structural information of XML documents under different granularities. We also address the issue of meaningful results clustering based on CCG and relevance oriented ranking strategy within clusters based on SCG. Besides, we define a novel semantics CLCA, which improves both the accuracy and completeness of search results. The experimental results indicates the efficiency and effectiveness of our proposed method.

References

1. Liu, Z., Chen, Y.: Identifying meaningful return information for xml keyword search. In: SIGMOD, pp. 329–340 (2007)
2. Liu, Z., Chen, Y.: Return specification inference and result clustering for keyword search on xml. ACM TODS 35(2), 1–47 (2010)
3. Yang, W., Zhu, H.: Semantic-distance based clustering for xml keyword search. In: Zaki, M.J., Yu, J.X., Ravindran, B., Pudi, V. (eds.) PAKDD 2010. LNCS, vol. 6119, pp. 398–409. Springer, Heidelberg (2010)
4. Liu, Z., Chen, Y.: Processing keyword search on xml: A survey. World Wide Web 14(5-6), 671–707 (2011)
5. Zhou, R., Liu, C., Li, J., Yu, J.X.: Elca evaluation for keyword search on probabilistic xml data. World Wide Web 16(2), 171–193 (2013)
6. Liu, X., Wan, C., Chen, L.: Returning clustered results for keyword search on xml documents. IEEE TKDE 23(12), 1811–1825 (2011)
7. Washington xml data repository,
   http://www.cs.washington.edu/research/xmldatasets/

Logics for Representing Data Mining Tasks in Inductive Databases

Hong-Cheu Liu, Millist Vincent, Jixue Liu, and Jiuyong Li

School of Information Technology and Mathematical Sciences
University of South Australia
Mawson Lakes Campus, Adelaide 5095, Australia
{Hong-Cheu.Liu,Millist.Vincent,Jixue.Liu,Jiuyong.Li}@unisa.edu.au

Abstract. We present a logical framework for querying inductive databases, which can accommodate a variety of data mining tasks, such as classification, clustering, finding frequent patterns and outliers detection. We also address the important issues of the expressive power of inductive query languages. We show that the proposed logic programming paradigm has equivalent expressive power to an algebra for data mining presented in the literature [1].

1 Introduction

The integration of data mining with the underlying databases systems has led to the development of the concept of an *inductive database* which, while originally proposed some years ago [2], remains an active research area [3,4]. The key idea of inductive database systems is that patterns (or models) are treated in the same way as data, i.e. as the first class citizens. In this context, an inductive database instance contains data, patterns, models and mixture of these objects [3]. In inductive database systems, the query language can be used to perform a number of tasks such as specifying constraints; retrieving, manipulating and applying interesting data, patterns, models; and matching data with specified patterns or models. The distinctive feature of an inductive database system compared to a general data mining system is that all knowledge discovery processes can be seen as an extension of the query session [2].

While the topic of data mining and knowledge discovery has made important advances over the past two decades [5], there is still a lack of a general formal framework for representing data mining and knowledge discovery tasks. [6,3]. This issue is well recognised, and a constant theme found in data mining literature is that the knowledge discovery processes should be supported by database technology, and that data mining methods should be integrated into a database system [3,7,2,8]. There is no doubt that the theoretical foundations and theory-directed system developments will have to advance significantly before it can be claimed that data mining is a self-contained and well developed *data science* similar to relational database theory.

Inductive databases will become increasingly important as the volumes of data increase, for reasons we now outline. The problem with many stand-alone

H. Wang and M.A. Sharaf (Eds.): ADC 2014, LNCS 8506, pp. 214–222, 2014.

data mining algorithms is that they do not scale to very large data sets, and so will not be useful in the coming era of 'big data' [9]. The advantage of inductive databases in this context is that since previously mined knowledge patterns are stored, these patterns can be incrementally updated when new data is added to the database rather than having to discover the pattern from scratch. In general, the incremental approach is much more efficient than recompilation from scratch. Also, for new data mining tasks posed to an inductive database, they can often be expressed as compositions and/or combinations of known mining tasks which are already stored in an inductive database. For example, an analyst might wish to find a collection of frequent item-sets purchased from a retail chain. The analyst may then combine these itemsets with a decision tree which classifies customers according to credit risk or income, which enables frequent itemsets to be further analysed according to customer type. Such composition of data mining tasks can be easily expressed in a logic programming framework, which is also amenable to optimisation.

The success of inductive databases will depend on whether they can efficiently and effectively mine complex and/or structured data/pattern/models and reason about query evaluation and optimisation. Several projects, based on ad hoc extensions to SQL, have made a number of contributions including exploring and demonstrating some of the key features required in an inductive DBMS [10,11]. However, these SQL-based extensions have a number of limitations. The first is that the SQL language was originally designed for the retrieval of data, not for storage of patterns and models, and so is ill suited to expressing many data mining tasks. The second limitation is that these extensions are ad hoc and they also lack a unifying formal framework in the way that relational databases are founded on the relational model. Finally, the SQL extensions have been of a declarative nature and the issue how to efficiently implement such extensions has not been addressed.

An alternative approach to developing inductive databases is based on logic programming. The advantage of this approach is that it is much easier to describe and compose data mining tasks in this framework, and also some of the work has addressed the issue how to effectively implement this logic-based approach. Also, because of the declarative nature of logic programming it is amenable to implementation and optimisation as already shown in logic programming languages such as Datalog. However, the limitation of previous work using this logic-based approach is that they are based on the flat relational model which makes them awkward to use in data mining tasks which inherently use more complex data such sequences and sets.

In this paper, we extend previous work on logic-based approach to inductive databases by proposing a new approach based on the complex value data model, as opposed to the flat relational model used in previous work. The motivation to doing this is to overcome the limitation just discussed of the flat relational approach in handling those data mining tasks which naturally involve complex data, such as sequences and sets.

In more detail, the main results of this paper are summarised as follows.

- We adopt a complex value data model for integrating data mining into databases which is easy to model semi-structured/complex structured data objects to be mined.
- We demonstrate that various data mining tasks including frequent pattern mining, decision tree induction, clustering and outliers detection application, can be expressed in our logic programming framework. The closure property of our language enables the composition of query results.
- We show that the expressive power of our proposed logic approach is equivalent to the data mining algebra presented in [1], which is slightly different to the nested relational algebra extended with aggregations proposed in [6]. As there is trade-off between the expressive power of queries and data complexity, in this article we only consider complex value calculus/algebra extended with aggregations and fixpoint operator, rather than Turing machine complete language. The proposed query class already can express most data mining tasks.

2   Logic Foundations for Inductive Queries

2.1   Deductive vs Inductive Databases

There is an analogy between deductive databases and inductive databases. They both are formalised in first-order (or Monadic second order) logic expressed in logic programming setting. However, there exists significant difference on semantics interpretation and data manipulation approach in these two types of databases. In deductive databases, queries produce new tuples (facts) from the known extensional database. For example, recall the classical logic program for computing the transitive closure of a graph G: $T(x,y) \leftarrow G(x,y)$, $T(x,y) \leftarrow G(x,z), T(z,y)$. This program produces all binary relations which are the transitive closure of G. The set of (x,y) already exist in the originally database.

In contrast, in inductive databases we integrate data mining with databases. In inductive databases we are more interested in finding new patterns and/or models and conducting data analysis in addition to data manipunation. For a simple example, suppose that we have a database $D = \{aa, bb, ab\}$. We apply an algorithm to generate a pattern view $v1 = freq(t, D) \geq 2$, where $t \in D$, $freq$ denotes the frequency of t occurring in D. At this point, $v1 = \{a, b\}$. When we update database D by inserting $\{abc\}$. Then $v1$ would be changed to $v1 \cup \{ab\}$. They not only store original data set but also store mined patterns/models.

2.2   Concept Learning

One of the most basic tasks in machine learning is *inductive concept learning* . The task of inductive concept learning is defined as follows.

Definition 1. *Given $\mathcal{U}$ and a concept $\mathcal{CP}$, find a hypothesis (classifier) H which is able to distinguish whether $x \in \mathcal{CP}$, for each $x \in \mathcal{U}$.*

A pattern class is a set of objects which satisfy some conditions (e.g., length, frequency, etc). Informally, let $\mathcal{D}$ be a data set, $\mathcal{P}$ a pattern class. $\mathcal{L_P}$ denotes the language describing the pattern class $\mathcal{P}$. A model $\mathcal{M}$ can be generated by some data mining task. An inductive database $\mathcal{I} = \mathcal{D} \cup \mathcal{P} \cup \mathcal{M}$.

Definition 2. *Given an inductive database instance $\mathcal{I}$ and a pattern discovery task can be specified as a query q such that $q(\mathcal{I}) = \{p \mid (\mathcal{I}, \mathcal{L_P}) \models p\}$,*

Definition 3. *Given an inductive database instance $\mathcal{I}$ and a data set $\mathcal{E}$, $\mathcal{E} \subset \mathcal{I}$, consisting of examples of pairs (x, y). x is of type T_x and y is of type T_y. A learning predictive model can be specified as a query $q_{\mathcal{E}}(\mathcal{I}) = \{\mathcal{M} \mid \mathcal{I} \models \mathcal{M}, (x, y) \in \mathcal{E}, (x, \hat{y}) \doteq \mathcal{M} \wedge y \simeq \hat{y}\}$, where $\mathcal{M}$ is a predictive model and $y, \hat{y}$ are observed and predicated values respectively. $\doteq$ denotes an example satisfying a model. $\simeq$ denotes that two values match closely.*

Definition 4. *Given an inductive database instance $\mathcal{I}$ and a set of examples $\mathcal{E}$, $\mathcal{E} \subset \mathcal{I}$. A clustering task can be specified as a query: $q_{\mathcal{E}}(\mathcal{I}) = \{C_i \mid \forall x \in C_i, y \in C_j, i \neq j, f_d(x, y) \geq c \text{ and } \forall x, y \in C_l, 1 \leq l \leq k, f_d(x, y) < c'\}$, where C_i are mined clusters and f_d is a distance metric function for measuring similarity between two objects. c and c' are threshold values.*

2.3 Constraint-based Mining

Constraint-based mining is a core technique used in inductive query evaluation since constraints can be used to confine the search space.

Definition 5. *A constraint is a predicate formula φ applied to a subset of the data set contained in an inductive database $\mathcal{I}$, $\varphi : Powerset(\mathcal{D}) \mapsto \{true, false\}$ or is a specific criterion which holds in $\mathcal{P}$ or $\mathcal{M}$.*

Example 1. Let $\mathcal{D}$ be a data set of an inductive databases with a language of patterns $\mathcal{L_P}$. An association rule is an element of the set $q(\mathcal{D}) = \{(A, B) \mid A, B \in \mathcal{L_P}, freq(A \cup B, \mathcal{D}) \geq s \text{ and } freq(A \cup B, \mathcal{D})/freq(A, \mathcal{D}) \geq c\}$. Where s is the support threshold and c is the confidence threshold. The query result (A, B) stands for an association rule $A \Rightarrow B$.

2.4 Fixpoint Operator

We define a fixpoint operator which allows the iteration of calculus formulas up to a fixpoint. This inductive mining loop operator is inspired by the the viewpoint that most data mining algorithms start out with an initial set of examples, possibly empty, then iteratively refine it until a termination condition is satisfied. The motivation for defining a fixpoint operator in a data mining query language is to provide an alternative way to achieve data mining tasks and to assist the development of a logic database language with data mining mechanisms for finding patterns, models extraction, obtaining both induced and deduced knowledge.

The inflationary version of the fixpoint operator is presented as follows. Let $\mathbf{R}$ be an inductive database schema, and T be a complex value relation schema not in $\mathbf{R}$. Let $\mathbf{S}$ denote the schema $\mathbf{R} \cup \{T\}$. Let $\varphi(T)$ be a formula using T and relations in $\mathbf{R}$, with some pattern mining *constraints*. Given an instance $\mathcal{I}$ over $\mathbf{R}$, $\mu_T(\varphi(T))$ denotes the relation that is the limit, if it exists, of the sequences $\{\lambda_n\}_{n \geq 0}$ defined by $\lambda_0 = \emptyset$; $\lambda_n = \lambda_{n-1} \cup \varphi(\lambda_{n-1}), n > 0$. This definition ensures that the sequence $\{\lambda_n\}_{n \geq 0}$ is increasing: $\lambda_{i-1} \subseteq \lambda_i$ for each $i > 0$. As there are finitely many tuples (items) that can be added, the sequence converges in all cases. That is, there exists some k for which $\lambda_k = \lambda_j$ for every $j > k$. Clearly, λ_k holds the set of patterns of $\mathbf{R}$. Note that $\lambda_k = \varphi(\lambda_k)$, so λ_k is also a *fixpoint* of $\varphi(T)$. The relation λ_k thereby obtained is denoted by $\mu_T(\varphi(T))$. By definition, μ_T is an operator that produces a new relation (the fixpoint λ_k) when applied to $\varphi(T)$.

Example 2. Let $D = (TID, List\_of\_item)$ be a transaction table. $List\_of\_item$ is a set valued attribute. The frequent itemset relations λ_n holding pairs of *itemset* and *count* can be defined inductively using the following formula

$$\mu_{T^i}[G = (\ _{itemset}\mathcal{G}_{count(TID)}(D) \wedge itemset \subset List\_of\_item \wedge |itemset| = i) \wedge G[count] \geq s \wedge T = G]$$

Where $\mathcal{G}$ denotes an aggregation function, μ_{T^i} denotes the fixpoint operater at the iteration step i.

3   A Logic Query Language

Like deductive databases, the notion of inductive databases can be specified in logic programming languages, such as Datalog. The key idea of the model-theoretic approach is to view the program as a set of higher-order sentences that describes the desired answer. Thus the inductive database instance consists of the mined patterns and models satisfying the sentences. Such an instance is also called a *model* of the sentences.

3.1   Frequent Pattern Mining

We illustrate an example for the frequent pattern mining task with some constraint.

Example 3. Let a transaction relation be $T = (ID, Items)$ and each item in the transaction database has an attribute value (such as profit) stored in a separate table $Profit$. The constraint $C_{avg} \equiv avg(S) \geq 25$ requires that for each itemset S, the average of the profits of the items in S must be equal or greater than 25. The frequent pattern mining task is to find all frequent itemsets such that the above constraint holds. We express it as inductive clauses as follows.

$cand(J, ID) \qquad\qquad \leftarrow T(ID, Items), J \subset Items$
$freq(J, count < ID >) \leftarrow cand(J, ID)$
$cfp(J, SUM < value >) \leftarrow freq(J, c), c > \delta, Profit(item, value), item \in J$
$Ans(Items) \qquad\qquad\quad \leftarrow cfp(J, SUM), avg = SUM/|J|, avg \geq 25$

3.2 Cluster Analysis: Partitioning Method

The k-*means* method takes the input parameter, k, and partitions a set of n objects into k clusters. We arbitrarily choose k objects from data as the initial cluster centers stored in table $cluster(c, \langle y \rangle, m)$, where $mean$ is a function used to calculate the cluster mean value. We use grouping construct mechnism adopted from the language $\mathcal{LDL}$. $\langle y \rangle$ is a set of values which belongs to class c.

For each of the remaining objects, an object is assigned to the cluster to which it is the most similar, based on the distance between the object and the cluster mean. It then computes the new mean for each cluster. Eventually, no redistribution of the objects in any cluster occurs and so the process terminates.

The following shows the clustering process.

Example 4. The clustering process can be expressed as follows.

$new\_cluster(c, \langle z \rangle) \leftarrow r(x), \ cluster(c, \langle y \rangle, m), \ min(distance(x, m)), \ ins(x, \langle y \rangle, \langle z \rangle)$
$new\_cluster(c, \langle y \rangle) \leftarrow new\_cluster(c, \langle z \rangle) \ , \ cluster(c, \langle y \rangle, m), \ x \in \{\langle z \rangle\} \cap \{\langle y \rangle\},$
$$delete(x, \langle y \rangle)$$
$cluster(c, \langle w \rangle, m) \ \leftarrow new\_cluster(c, \langle w \rangle) \ , \ m = mean\{\langle w \rangle\}$

$distance$ is a similarity function. $ins(x, \langle y \rangle, \langle z \rangle)$ is interpreted as "set $\langle z \rangle$ is obtained by inserting x into $\langle y \rangle$".

3.3 Decision Tree Induction

In this subsection, we give a general framework of how to formulate a logic program for modeling decision tree induction. For simplicity, we assume that the input of the problem is a relation $D(A_1, ..., A_n, CL)$. The attribute CL identifies the class the tuple belongs to. The output will be a relation $\mathcal{T} = (branch, Label)$ that holds the set of all inequalities associated with the leaves of the tree, together with their labels. That is, for each leaf in the tree we generate the set of all inequalities on the edges from the root to that leaf. For example, suppose that there is a leaf in the tree which is reached via edges labeled, respectively, age $<$ 25, student $=$ 'yes', and having class label 'yes'. For this leaf, the relation $\mathcal{T}$ will contain the tuple $< branch : \{age < 25, student = 'yes'\}, Label : 'yes' >$.

This approach is different from the one adopted by previous work of [12] as we use the complex value data model and adopt fixpoint semantic approach.

3.4 Application: Outliers Detection

In this subsection, we briefly describe that outlier detection can be formalised in logic programming paradigm by using default logic [13,14].

The major difference between our proposal stated in this subsection and previous works is that we consider the realm of first-order and monadic second order default logic by adopting fix-point semantics. The reason is that finding outliers is quite complex and a propositional default theory may not be sufficient.

Default logics. A default theory $\mathcal{T}$ is a pair $(\mathcal{R}, \mathcal{F})$ consisting of a set $\mathcal{F}$ of firs-order or monadic second order logic formulas and a set $\mathcal{R}$ of default rules. A default rule γ has the form $\frac{\rho : \varphi_1, ..., \varphi_k}{\lambda}$, where ρ, φ_i, and γ are first order or monadic second order formulas. ρ is called the prerequisite, φ_i the justification, and λ the conclusion of γ. The informal meaning of a default rule γ is the following: if prerequisite ρ is known to hold, and if it is consistent to the justification φ_i, then we get conclusion γ.

The semantics of a default theory is defined as a fixpoint of a Datalog program formed by using prerequisites, justifications and conclusions. That is:

$$J_0 = \mathrm{ground}(\mathcal{F}); \; J_n = J_{n-1}^* \cup \varphi(J_{n-1}), \, n > 0$$

where J_{n-1}^* is the closure of J_{n-1}, $\varphi(J_{n-1}) = \{\lambda \mid \frac{\rho : \varphi_1, \ldots, \varphi_k}{\lambda} \in \mathcal{R}\}$. The semantics of $\mathcal{T}$ is a fixpoint $\mu_T(\varphi(J_n))$.

Definition 6. *Let* $\mathcal{T} = (\mathcal{R}, \mathcal{F})$ *be a default theory and let* $l \in \mathcal{F}$ *be a literal. If there exists a non-empty set of literals* $L \subseteq \mathcal{F}$ *such that* $(\mathcal{R}, \mathcal{F}) \models \neg L$, *and* $(\mathcal{R}, \mathcal{F} - \{l, L\}) \models \neg L$ *does not hold.*

Then we say that l is an outlier in $\mathcal{T}$ and L is an outlier witness set for l in $\mathcal{T}$.

4   Expressive Power

We investigate the expressive power of our logic-based calculus as follows.

Theorem 1. *Any data mining queries expressible in logic-based calculus with mining loop can be specified as inductive clauses in* $Datalog^{cv,\neg}$.

PROOF SKETCH. A query is expressible in $Datalog^{cv,\neg}$ with stratified negation if and only if it is expressible in complex value calculus $CALC^{cv}$. $CALC^{cv}$ is equivalent to $CALC^{cv}$ + fixpoint. So $Datalog^{cv,\neg}$ with stratified negation is equivalent to $CALC^{cv}$ + fix-point. Any data mining queries expressible in logic-based calculus with mining loop can be specified as inductive clauses in $Datalog^{cv,\neg}$. $\square$

One major factor for promising inductive database to be successful is the formulation of an 'algebra' for query optimisation and reasoning. We briefly describe a data mining algebra which was proposed in [1] as follows.

4.1   A Data Mining Algebra

Let $\Omega = \{R_1, ..., R_n\}$ be a signature, where R_i, $1 \leq i \leq n$, are database relations. The data mining algebra over Ω is denoted as $\mathcal{DMA}(\Omega)$. A family of core operators of the algebra is presented as follows. **Set operations:** Union ($\cup$), Cartesian product ($\times$), and difference (-) are binary set operations. **Tuple operations:** Selection (σ) and projection (π) are defined in the natural manner. **Powerset:** $powerset(r)$ is a relation of sort $\{\tau\}$ where $powerset(r) = \{\nu \mid \nu \subseteq r\}$. **Tuple Creation:** If $A_1, ..., A_n$ are distinct attributes, $tup\_create_{A_1,...,A_n}(r_1, ..., r_n)$ is of sort $< A_1 : \tau_1, ..., A_n : \tau_n >$, and $tup\_create_{A_1,...,A_n}(r_1, ..., r_n) = \{< A_1 : \nu_1, ..., A_n : \nu_n > \mid \forall i(\nu_i \in r_i)\}$. **Set Creation:** $set\_create(r)$ is of sort $\{\tau\}$, and $set\_create(r) = \{r\}$. **Tuple Destroy:** If r is of sort $< A : \tau' >$, $tup\_destroy(r)$ is a relation of sort τ' and $tup\_destroy(r) = \{\nu \mid < A : \nu > \in r\}$.

Set Destroy: If $\tau = \{\tau'\}$, then $set\_destroy(r)$ is a relation of sort τ' and $set\_destroy(r) = \cup r = \{w \mid \exists \nu \in r, w \in \nu\}$. **Aggregation:** The standard set of aggregate functions SUM, COUNT, AVG, MIN, MAX are defined in the usual manner. For example, if r is of sort $< A : \tau_1, B : \tau_2 >$, $\mathcal{G}_{<A>}^{function<B>}(r)$ is the relation over $< A, S >$. $\mathcal{G}_{<A>}^{function<B>}(r) = \{< a, s > \mid \exists < a, v > \in r \wedge s = \Sigma\{t < B > \mid t \in r, t < A, B >=< a, b >\}$, where Σ is one aggregate operator.

Theorem 2. *The expressive power of logic-based calculus with fixpoint operator for inductive queries is equivalent to the data mining algebra $\mathcal{DMA}$.*

PROOF SKETCH. In the complex value data model, $CALC^{cv} + \mu$ is equivalent to the algebra ALG^{cv}. The aggregation operators can be expressed by composition of primitive algebraic operators. Therefore the expressive power of logic-based calculus with fixpoint operator for inductive queries is equivalent to the data mining algebra $\mathcal{DMA}$. □

5 Conclusion

We have presented a logical framework for querying inductive databases. The framework would be helpful for understanding querying aspects of inductive databases. We have also presented an inductive logic programming query language and illustrated the use of aggregates and exploit a fixpoint operator to model specific data mining tasks. The results provide theoretical foundations for inductive database research and could be useful for query language design in inductive database systems.

Acknowledgment. This work was supported by the grant DP1096523 from the Australian Research Council.

References

1. Liu, H.-C., Ghose, A., Zeleznikow, J.: Towards an algebraic framework for querying inductive databases. In: Kitagawa, H., Ishikawa, Y., Li, Q., Watanabe, C. (eds.) DASFAA 2010, Part II. LNCS, vol. 5982, pp. 306–312. Springer, Heidelberg (2010)
2. Imielinski, T., Mannila, H.: A database perspective on knowledge discovery. Communications of the ACM 39(11), 58–64 (1996)
3. Džeroski, S.: Inductive databases and constraint-based data mining. In: Jäschke, R. (ed.) ICFCA 2011. LNCS (LNAI), vol. 6628, pp. 1–17. Springer, Heidelberg (2011)
4. Romei, A., Turini, F.: Inductive databases languages: Requirements and examples. Knowledge Information Systems 26, 351–384 (2011)
5. Han, J.: Data Mining: Concepts and Techniques. Morgan Kaufmann (2000)
6. Calders, T., Lakshmanan, L., Ng, R., Paredaens, J.: Expressive power of an algebra for data mining. ACM Transactions on Database Systems 31(4), 1169–1214 (2006)

7. Giannotti, F., Manco, G., Turini, F.: Towards a logic query language for data mining. In: Meo, R., Lanzi, P.L., Klemettinen, M. (eds.) Database Support for Data Mining Applications. LNCS (LNAI), vol. 2682, pp. 76–94. Springer, Heidelberg (2004)
8. Raedt, L.D.: A perspective on inductive databases. SIGKDD Explorations 4(2), 69–77 (2002)
9. Mayer-Schonberger, V., Cukier, K.: Big Data: A Revolution That Will Transform How We Live, Work, and Think. Eamon Dolan/Houghton Mifflin Harcourt (2013)
10. Han, J., Fu, Y., Koperski, K., Wang, W., Zaiane, O.: Dmql: A data mining query language for relational databases. In: Proceedings of ACM SIGMOD Workshop on Research Issues on Data Mining and Knowledge Discovery (1996)
11. Meo, R., Psaila, G., Ceri, S.: An extension to sql for mining association rules. Data Mining and Knowledge Discovery 2(2), 195–224 (1998)
12. Nijssen, S., De Raedt, L.: Iql: A proposal for an inductive query language. In: Džeroski, S., Struyf, J. (eds.) KDID 2006. LNCS, vol. 4747, pp. 189–207. Springer, Heidelberg (2007)
13. Angiulli, F., Ben-Eliyahu-Zohary, R., Palopoli, L.: Outlier detection using default reasoning. Artificial Intelligence, Elsevier 172, 1837–1872 (2008)
14. Reiter, R.: A logic for default reasoning. Artificial Intelligence 13(1-2), 81–132 (1980)

An Effective Approach to Handling Noise and Drift in Electronic Noses

Sanad Al-Maskari*, Xue Li, and Qihe Liu

School of Information Technology and Electrical Engineering
The University of Queensland, Australia
Sohar University, Environmental Research Centre, Sohar PC 311, Sultanate of Oman
s.almaskari@uq.edu.au, xueli@itee.uq.edu.au, qiheliu@uestc.edu.cn

Abstract. Sensor drift and noise handling in electronic noses (E-noses) are two different challenging problems. Sensor noise is caused by many factors such as temperature, pressure, humidity and cross interference. Noise can occur at any time, producing irrelevant or meaningless data. On the other hand, drift is appeared as a long term signal variation caused by unknown dynamic physical and chemical complex processes. Because sensor drift is not purely deterministic, it is very hard, if not impossible, to distinguish it from noise and vice versa. With respect to this property of E-nose, we propose a new approach to handle noise and sensor drift simultaneously. Our approach is based on kernel Fuzzy C-Mean clustering and fuzzy SVM (K-FSVM). The proposed method is compared to other currently used approaches, SVM, F-FSVM and KNN. Experiments are conducted on publicly available datasets. As the experimental results demonstrate, the performance of our proposed K-FSVM is superior than all other baseline methods in handling sensor drift and noise.

Keywords: E-nose, noise, drift , classification, Kernel Fuzzy C-Mean.

1 Introduction

Electronic nose (E-nose) is a generic name for devices capable of identifying chemical analytes or measuring odour information in different environments [1]. E-nose systems have been used in a wide range of applications including: food industry, agriculture, air quality and environment monitoring , odour monitoring in a poultry shed, medicine , water and waste water quality control [2,1,3]. Despite their reputation, their practical values are affected by their poor stability making them very vulnerable to drift and noise. These vulnerabilities can be resolved using periodic recalibration which in turn increases their cost significantly and make them overly complex. Eliminating the need for periodic recalibration and the ability to extend the recalibration period using machine learning will result in resolving the main issues with these sensors.

---

* Corresponding author.

H. Wang and M.A. Sharaf (Eds.): ADC 2014, LNCS 8506, pp. 223–230, 2014.
© Springer International Publishing Switzerland 2014

Sensor drift is the gradual and unpredictable variation of the chemo-sensory signal responses when exposed to the same analyte under identical conditions [4,5]. Such variations could be due to sensor poisoning, ageing and environmental changes such as humidity, temperature, pressure and system sampling non-specific adsorption [6]. In the other hand **noise** can be defined as any unwanted effect that obscures the detection measurement of the desired signal. Large noise can be generated by E-nose due to system circuit faults, sensor poisoning, or ageing, and environmental effects. Because sensor drift is not purely deterministic, it is very hard, if not impossible, to distinguish it from noise and vice versa [7,8].

In this paper we handle sensor drift and noise as an abnormal phenomenon that has two different indicators mixed together. Our approach introduces a new perspective on how E-nose sensor drift and noise should be handled. To the best of our knowledge, this approach has not been applied in the chemical sensing community before. In our approach, a classifier with noise and drift resistant based on Fuzzy SVM is proposed. In our proposed approach, two different member generation methods (FCM and KFCM) are used and compared. Furthermore, an optimized Kernel based fuzzy clustering algorithm is proposed. The experiments based on the UCI data set are conducted and show that our proposed approach outperforms all other baseline approaches currently used in E-nose data processing.

This paper is organized as follows. Section 2 gives a comprehensive overview on the algorithms for E-nose data processing. Section 3 describes the proposed approach. Section 4 gives the details of the experiments and the evaluation of the results. Section 5 concludes our paper.

2   Related Work

Univariate and Multivariate methods are commonly used to deal with sensor drift. Univariate methods are used to measure central tendency, frequency distribution, and analyse each variable output pattern. Despite their simplicity and their low computation complexity they fail to capture complex and non-linear drift or correlated drift effects and they required periodic recalibration [9]. One of the best univariate methods (univariate multiplicative factor) introduced in [9] requires recalibration every 3 months. Various multivariate methods have been proposed to deal with sensor drift such as PLS, PCA, CCA. In some studies PCA is used to find the directions of maximum variance of data x whereas CCA identifies directions where two variables x, y co-vary. In [10] a supervised Orthogonal Signal Correction (OSC) method is proposed and found to perform better than unsupervised PCA-CC for the first 100 days. All these methods assume linear drift direction; where in real environment, E-noses are found to be strongly non-linear which makes them unable to handle multiple drift directions [11,9]. Unsupervised and supervised adaptive methods such as Self Organizing Maps (SOMs), multiple SOM , Neural networks, local class-dependent drift estimation also proposed to handle sensor drift [12,5,13]. These methods do not provide a generic and dynamic model capable of handling complex and dynamic environmental changes due to their sensitivity to noise and over fitting.

3 Proposed Method

Since data generated by E-nose are uncertain and unstable, a fuzzy approach is proposed to handle such issues. When a gas mixture is detected by E-nose unknown responses can be generated and such responses could belong to different classes with different percentages. Therefore, some data points could have different levels of importance or relevance. The standard SVM is unable to handle unknown or uncertain output generated by E-nose sensor due to its sensitivity to noise and outlier. Furthermore, standard SVM has only one free parameter (C) whereas in FSVM there are N possible free parameters (N is the total number of training points). A degree of importance s_i is given to each data point providing a greater flexibility and generalization to the model. For each training pair (x_i, y_i) a membership value s_i is given, the pairs with high s_i values will have a greater influence in the decision surface compared to the one with lower s_i values. In this method we use a Kernel-based Fuzzy C-Means (K-FCM) clustering algorithm to generate membership for our fuzzy classifier [14].

3.1 Clustering Process

The first step in our approach is to cluster the E-nose data sets. The idea is to group similar sensor features within the same partition by clustering. Once the optimal groupings of sensor features are found, the membership can then be used by the fuzzy classifier. Different features can have different types of information and calculating the total importance of each signal from the feature set is very critical for improving the learning model. To achieve this goal we employ a *fuzzy clustering* approach which enables us to create a relationship matrix between each signal feature set and its group. Unlike hard clustering methods, in fuzzy clustering one object can belong to different clusters with different membership degrees. One of the most well-known fuzzy clustering algorithms is Fuzzy C-Mean (FCM). FCM uses Euclidean norm to measure similarity between data points making it effective in handling spherical clusters, but it is sensitive to noise, outliers [15,14]. Kernel-based methods make it possible to handle tasks using a richer framework than the linear ones. Kernel-based FCM was introduced to overcome noise and outliers' sensitivity found in FCM by transforming input space X to a high or infinite dimension feature F space ($\phi : X \rightarrow F$). A Kernel-based Fuzzy C-Means clustering (KFCM) algorithm has been proposed by Zhang *et al.* [14]. KFCM partitions a given data set $X = \{x_i, ..., x_n\} \in R^p$ into C fuzzy subsets by minimizing the following objective function:

$$J_m(U, V) = \sum_{i=1}^{c} \sum_{k=1}^{n} u_{ik}^m ||\phi(X_K) - \phi(V_i)||^2 \qquad (1)$$

S.T: $\displaystyle\sum_{k}^{n} u_{ik} > 0, \forall i \in 1, ...c$ (2) $\displaystyle\sum_{i}^{c} u_{ik} > 1, \forall k \in 1, ...n$ (3)

Where, c is the number of clusters ($1 < c < n$). n is the number of data points; u_{ik} is the membership of X_k in class i satisfying $\sum_{i}^{c} u_{ik} = 1$ for all k and $u_{ik} \in [0, 1]$; m amount of fuzziness ($m > 1$); V is set of control cluster centres ($V_i \in^p$); ϕ is an implicit nonlinear transformation function. The Euclidean distance between points and centres in the feature space F can be computed as,

$$\|\phi(X_K) - \phi(V_i)\|^2 = k(X_K, X_K) + k(V_i, V_i) - 2k(X_K, V_i) \tag{4}$$

Because $K(x,x) = 1$ the Gaussian Kernel leads to $d\phi^2(x,y) = K(x,x) + K(y,y) - 2K(x,y) = 2(1 - K(x,y))$. Thus equation 1 becomes:

$$J_m(U,V) = 2 \sum_{i=1}^{c} \sum_{k=1}^{n} u_{ik}^m (1 = k(X_K, V_i)) \tag{5}$$

where,
$$k(X_K, V_i) = exp(-\|X_K - V_i\|^2/\sigma^2) \tag{6}$$

The optimization problem is solved by minimizing $J_m(U,V)$ under the constraints of u_{ik}.

$$u_{ik} = \frac{(1/(1-k(X_K,V_i)))^{1/(m-1)}}{\sum_{j=1}^{c}(1/(1-k(X_K,V_i)))^{1/(m-1)}}, \tag{9} \quad v_i = \frac{\sum_{k=1}^{n} u_{ik}^m K(X_K,V_i)X_K}{\sum_{k=1}^{n} u_{ik}^m K(X_K,V_i)} \tag{10}$$
$$\forall i \in 1...c \text{ and } \forall k \in 1...n$$

A kernel-based FCM algorithm is used to generate membership values for each data point X_i in the E-nose data set.Algorithm details are as follows:

Algorithm 1. Kernel Fuzzy C-Mean Clustering Algorithm

Input:
    b_n:Data batches $b_1, b_2,...b_n$
    m:Fuzzification parameter
    C:Number of clusters
    ε :set termination parameter;
Output: optimal member ship matrix U_{opt}
1. Select the kernel function K and its parameters;
2. Compute Kernel Values;
3. Initialize membership matrix ui;
4. Select cluster centers vi;
5. Update membership matrix u_{ik} using equation 9 ;
6. Compute all new cluster centers or prototype v_i using equation 10;
7. Repeat step 5-6 and check the termination function E^t;
    $E^t = max|U_new - U_old|$, if $E_t \leq \varepsilon$, stop;
8. repeat step 3-7 until optimal membership matrix is found;
9. Select and save optimal membership matrix U_{opt};

3.2   Fuzzy Classifier

SVM is widely regarded as one of the most powerful tools used for solving classification problems. However, it is prone to over fitting which makes it very sensitive to data sets with noise. In standard SVM, a training point can belongs to either one class or another. Most real world applications would have some training points with higher importance than others. Consequently, the training points with higher importance should be classified correctly and the noisy points or meaningless ones will not be considered and therefore discarded. Fuzzy SVM allows each data point X_i to be assigned a membership value U_i where $0 < s_i \leq 1$. The membership s_i is used to determine the importance or relativity of each data point X_i to one class and the value $1 - s_i$ can be used to determine the degree of meaningless. Memberships is generated using FCM and KFCM. Using one-vs-one strategy (OVO) a Fuzzy SVM model is constructed. Better performance and generalization is expected when combining membership with FSVM model. The optimal training model is obtained by searching through a grid of values to find best γ and C $[2^{-10}, 2^{-9}, .., 2^4, 2^5]$ and $[2^{-5}, 2^{-4}, .., 2^9, 2^{10}]$ [5]. The following steps are performed to execute the K-FSVM model:

Algorithm 2. Fuzzy SVM Classification algorithm

Input:
    b_n:Data batches $b_1, b_2, ... b_n$
    U_opt: Membership matrix U_opt from algorithm 1
    σ , γ and ε Parameter
Output: final prediction matrix
1.   Use OVO strategy to create multiple classifiers;
2.   Generate the relationship matrix R_n from U_opt ;
3.   From membership matrix R_n train FSVM using $\{x_i, y_i, R_{ni} \}$.
4.   predict all class labels using voting
5.   **return** final classifier;

---

4   Experiments and Evaluation

The data set used in this paper is presented by Vergara *et al.* [5]. In their three-year experiment they collected an extensive data set using metal-oxide gas sensor array for six-gas/analyte. The gas was dosed at different concentrations and their goal was to identify different analytes type regardless of their concentrations. The experiment was designed to emulate sensor drift and it was shown by Vergara *et al.* that sensor drift actually occurred. The final data set contains 13,910 samples collected over 36 months. For more details about the dataset refer to [5]. The experiments are divided into two parts. The first part handles only sensor drift and the second part handles sensor drift and noise simultaneously. In our experiments two settings are considered:

    Setting 1: classifier trained with data from only the previous batch (b-1) and tested in the current batch b.

    Setting 2: classifier is trained on batch1 and tested on the remaining batches. Training SVM under setting 1 provides a strong baseline because the training batch is close to the testing batch which minimizes drift quantity. In order for drift to occur, it requires more time therefore setting 2 will demonstrate the classifier ability to handle drift.

4.1   Sensor Drift Handling

In this experiment K-FSVM with RBF kernel is trained to handle sensor drift on E-nose data set under *Setting 1* and *Setting 2*. We compared K-FSVM with F-FSVM and SVM. In this experiment the noise is not considered. Figures 1.0 and 2.0 shows the classification accuracies generated by SVM,F-SVM and K-SVM under Setting 1 and Setting 2. Figure 1 shows that classifiers trained under *Setting 1* performed better than classifiers trained under *Setting 2* as expected. The average classification accuracy of K-FSVM (82.18%) under setting 1 is better than other classifiers. The performance of SVM and KFSVM is comparable at batch 4,7, and 8 under setting 1 indicating that K-FSVM will not always perform better than SVM in cases with minimal drift. K-SVM and F-SVM performed equally at batch 4, 5, 6 and 8. Also F-SVM performed better than SVM in most instances except (7, 8, and 10). Since F-FSVM and K-FSVM work with a very similar concept in generating memberships they can provide comparable results in instances with low sensor drift. This indicates that K-FSVM will not always perform better than F-SVM in cases with low sensor drift. Overall K-FSVM performed better than F-FSVM.

The results of the experiment under *Setting 2* demonstrate sensor drift and can be observed clearly in Figure 2. The performances of all classifiers are degrading over time due to drift. In all cases K-FSVM has performed better than SVM classifier with average classification accuracy of 58.419%. Although K-FSVM has performed better than other methods in all batches, the prediction accuracies for batches 7, 8, 9, and 10 are less than 50% indicating the need for more improvements. Clearly using our proposed approach can improve the performance of sensor drift but it cannot eliminate it. There are still significant windows of enhancements to be achieved by improving our approach. Since sensor drift is

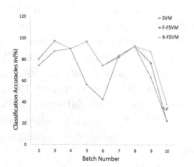

Fig. 1. The performance of K-FSVM under setting 1 compared with other classifiers

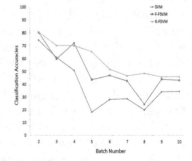

Fig. 2. The performance of K-FSVM under setting 2 compared with other classifiers

not a deterministic phenomenon, we believe a Fuzzy approach is a promising approach for handling sensor drift.

4.2   Sensor Drift and Noise Handling

In this experiment, classifiers of K-FSVM, F-FSVM, KNN and SVM with RBF kernel is trained to handle sensor drift on E-nose data set under *Setting 1*, *Setting 2*. Noise was introduced to the E-nose dataset to compare the abilities of the classifiers in handling sensor drift and noise. Above 15% of normally distributed additive white Gaussian Noise was added to the data set. Figures 3.0 and 4.0 shows the classification accuracies generated by SVM, F-SVM, K-SVM, and KNN under *Setting 1* and *Setting 2*. In both settings K-FSVM has performed better than other classifiers. Under *Setting 1* SVM has performed better only in two batches (4 and 5) and KNN performed best in batch 10. In batch 10 KNN performed better than K-SVM with 1.06% difference and SVM overtook K-FSVM with only 0.57% in the last batch. The ability of the fuzzy classifier to handle noise and sensor drift is demonstrated clearly in Figure 4.0. Under *Setting 2* K-FSVM has performed much better than other classifiers. While injecting over than 15% noise to dataset, K-FSVM only dropped 3.61% scoring average classification accuracy of 54.77% while F-KFSVM dropped over than 9.28% scoring 41.59%. All classifiers performed poorly on batches 7 to 10 which indicate the requirement for more improvements.

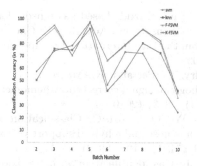

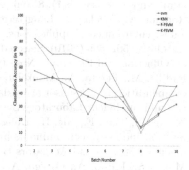

Fig. 3. The performance of all classifiers under setting 1

Fig. 4. The performance of all classifiers under setting 2

5   Conclusions

In this paper, a novel approach based on Fuzzy SVM is proposed and a fuzzy membership function is introduced. We propose using Kernel Fuzzy C-Mean to impose the membership into FSVM, enabling each of data points to be differentiated for its contribution to the learning of the decision surface. Different member generation methods (KFCM and FCM) are used and compared. The results show that K-FSVM provides a superior performance in handling sensor noise and drift when compared to other will-known baseline classifiers.

References

1. AL-Maskari, S., Saini, D.O.W.: Cyber infrastructure and data quality for environmental pollution control in Oman. In: Proceedings of the 2010 DAMD International Conference on Data Analysis, Data Quality and Metada, p. 71 (2010)
2. Nimsuk, N., Nakamoto, T.: Study on the odor classification in dynamical concentration robust against humidity and temperature changes. Sensors and Actuators B: Chemical 134(1), 252–257 (2008)
3. Pan, L., Yang, S.X.: A new electronic nose for downwind livestock farm odour measurement. In: Proceedings of the 2006 IEEE International Conference on Networking, Sensing and Control, ICNSC 2006, pp. 410–415 (2006)
4. Vergara, A., Vembu, S., Ayhan, T., Ryan, M.A., Homer, M.L., Huerta, R.: Chemical gas sensor drift compensation using classifier ensembles. Sensors and Actuators B: Chemical 166-167, 320–329 (2012)
5. Zuppa, M., Distante, C., Siciliano, P., Persaud, K.C.: Drift counteraction with multiple self-organising maps for an electronic nose. Sensors and Actuators B: Chemical 98(2-3), 305–317 (2004)
6. Sharma, R., Chan, P., Tang, Z., Yan, G., Hsing, I., Sin, J.: Investigation of stability and reliability of tin oxide thin-film for integrated micro-machined gas sensor devices. Sensors and Actuators B: Chemical 81(1), 9–16 (2001)
7. Goodner, K.L., Dreher, J., Rouseff, R.L.: The dangers of creating false classifications due to noise in electronic nose and similar multivariate analyses. Sensors and Actuators B: Chemical 80(3), 261–266 (2001)
8. Tian, F., Yang, S.X., Dong, K.: Circuit and noise analysis of odorant gas sensors in an e-nose. Sensors 5(1), 85–96 (2005)
9. Romain, A., Nicolas, J.: Long term stability of metal oxide-based gas sensors for e-nose environmental applications: An overview. Sensors and Actuators B: Chemical 146(2), 502–506 (2010); Selected papers from the 13th International Symposium on Olfaction and Electronic Nose ISOEN 2009
10. Padilla, M., Perera, A., Montoliu, I., Chaudry, A., Persaud, K., Marco, S.: Drift compensation of gas sensor array data by orthogonal signal correction. Chemometrics and Intelligent Laboratory Systems 100(1), 28–35 (2010)
11. Zhang, L., Tian, F., Nie, H., Dang, L., Li, G., Ye, Q., Kadri, C.: Classification of multiple indoor air contaminants by an electronic nose and a hybrid support vector machine. Sensors and Actuators B: Chemical 174, 114–125 (2012)
12. Marco, S., Pardo, A., Ortega, A., Samitier, J.: Gas identification with tin oxide sensor array and self organizing maps: Adaptive correction of sensor drifts. In: Instrumentation and Measurement Technology Conference, IMTC 1997, Proceedings of the Sensing, Processing, Networking, vol. 2, pp. 904–907. IEEE (May 1997)
13. Bishop, C.M.: Neural Networks for Pattern Recognition. Oxford University Press, Inc., New York (1995)
14. Zhang, D.Q., Chen, S.C.: A novel kernelized fuzzy c-means algorithm with application in medical image segmentation. Artificial Intelligence in Medicine 32(1), 37–50 (2004) (in China)
15. Krishnapuram, R., Keller, J.: A possibilistic approach to clustering. IEEE Transactions on Fuzzy Systems 1(2), 98–110 (1993)

Author Index